企业级卓越人才培养解决方案"十三五"规划教材

Android 模块化项目实战

天津滨海迅腾科技集团有限公司　主编

南开大学出版社

天　津

图书在版编目 (CIP) 数据

Android 模块化项目实战 / 天津滨海迅腾科技集团有限公司主编 . — 天津 : 南开大学出版社 , 2018.7
ISBN 978-7-310-05616-3

Ⅰ . ①A… Ⅱ . ①天… Ⅲ. ①移动终端－应用程序－程序设计 Ⅳ. ①TN929.53

中国版本图书馆 CIP 数据核字 (2018) 第 131993 号

主 编 归达伟 王 磊
副主编 熊祖涛 赵 杰 高德梅
周青政 马高平

版权所有 侵权必究

南开大学出版社出版发行
出版人 : 刘运峰
地址 : 天津市南开区卫津路 94 号 邮政编码 : 300071
营销部电话 : (022)23508339 23500755
营销部传真 : (022)23508542 邮购部电话 : (022)23502200

*

唐山鼎瑞印刷有限公司印刷
全国各地新华书店经销

*

2018 年 7 月第 1 版 2018 年 7 月第 1 次印刷
260×185 毫米 16 开本 14.75 印张 367 千字
定价 : 66.00 元

如遇图书印装质量问题 , 请与本社营销部联系调换 , 电话 : (022)23507125

企业级卓越人才培养解决方案"十三五"规划教材
编写委员会

指导专家： 周凤华　教育部职业技术教育中心研究所
　　　　　　李　伟　中国科学院计算技术研究所
　　　　　　张齐勋　北京大学
　　　　　　朱耀庭　南开大学
　　　　　　潘海生　天津大学
　　　　　　董永峰　河北工业大学
　　　　　　邓　蓓　天津中德应用技术大学
　　　　　　许世杰　中国职业技术教育网
　　　　　　郭红旗　天津软件行业协会
　　　　　　周　鹏　天津市工业和信息化委员会教育中心
　　　　　　邵荣强　天津滨海迅腾科技集团有限公司
主任委员： 王新强　天津中德应用技术大学
副主任委员： 张景强　天津职业大学
　　　　　　宋国庆　天津电子信息职业技术学院
　　　　　　闫　坤　天津机电职业技术学院
　　　　　　刘　胜　天津城市职业学院
　　　　　　郭社军　河北交通职业技术学院
　　　　　　刘少坤　河北工业职业技术学院
　　　　　　麻士琦　衡水职业技术学院
　　　　　　尹立云　宣化科技职业学院
　　　　　　廉新宇　唐山工业职业技术学院
　　　　　　张　捷　唐山科技职业技术学院
　　　　　　杜树宇　山东铝业职业学院
　　　　　　张　晖　山东药品食品职业学院
　　　　　　梁菊红　山东轻工职业学院
　　　　　　赵红军　山东工业职业学院
　　　　　　祝瑞玲　山东传媒职业学院
　　　　　　王建国　烟台黄金职业学院

陈章侠　德州职业技术学院
郑开阳　枣庄职业学院
张洪忠　临沂职业学院
常中华　青岛职业技术学院
刘月红　晋中职业技术学院
赵　娟　山西旅游职业学院
陈　炯　山西职业技术学院
陈怀玉　山西经贸职业学院
范文涵　山西财贸职业技术学院
郭长庚　许昌职业技术学院
许国强　湖南有色金属职业技术学院
孙　刚　南京信息职业技术学院
张雅珍　陕西工商职业学院
王国强　甘肃交通职业技术学院
周仲文　四川广播电视大学
杨志超　四川华新现代职业学院
董新民　安徽国际商务职业学院
谭维奇　安庆职业技术学院
张　燕　南开大学出版社

企业级卓越人才培养解决方案简介

企业级卓越人才培养解决方案（以下简称"解决方案"）是面向我国职业教育量身定制的应用型、技术技能型人才培养解决方案，以教育部 - 滨海迅腾科技集团产学合作协同育人项目为依托，依靠集团研发实力，联合国内职业教育领域相关政策研究机构、行业、企业、职业院校共同研究与实践的科研成果。本解决方案坚持"创新校企融合协同育人，推进校企合作模式改革"的宗旨，消化吸收德国"双元制"应用型人才培养模式，深入践行"基于工作过程"的技术技能型人才培养，设立工程实践创新培养的企业化培养解决方案。在服务国家战略，京津冀教育协同发展、中国制造 2025（工业信息化）等领域培养不同层次的技术技能人才，为推进我国实现教育现代化发挥积极作用。

该解决方案由"初、中、高级工程师"三个阶段构成，包含技术技能人才培养方案、专业教程、课程标准、数字资源包（标准课程包、企业项目包）、考评体系、认证体系、教学管理体系、就业管理体系等于一体。采用校企融合、产学融合、师资融合的模式在高校内共建大数据学院、虚拟现实技术学院、电子商务学院、艺术设计学院、互联网学院、软件学院、智慧物流学院、智能制造学院、工程师培养基地的方式，开展"卓越工程师培养计划"，开设系列"卓越工程师班"，"将企业人才需求标准、工作流程、研发项目、考评体系、一线工程师、准职业人才培养体系、企业管理体系引进课堂"，充分发挥校企双方特长，推动校企、校际合作，促进区域优质资源共建共享，实现卓越人才培养目标，达到企业人才培养及招录的标准。本解决方案已在全国近几十所高校开始实施，目前已形成企业、高校、学生三方共赢格局。未来三年将在 100 所以上高校实施，实现每年培养学生规模达到五万人以上。

天津滨海迅腾科技集团有限公司创建于 2008 年，是以 IT 产业为主导的高科技企业集团。集团业务范围已覆盖信息化集成、软件研发、职业教育、电子商务、互联网服务、生物科技、健康产业、日化产业等。集团以产业为背景，与高校共同开展产教融合、校企合作，培养了一批批互联网行业应用型技术人才，并吸纳大批毕业生加入集团，打造了以博士、硕士、企业一线工程师为主导的科研团队。集团先后荣获：天津市"五一"劳动奖状先进集体，天津市政府授予"AAA"级劳动关系和谐企业，天津市"文明单位"，天津市"工人先锋号"，天津市"青年文明号""功勋企业""科技小巨人企业""高科技型领军企业"等近百项荣誉。

前　言

在移动互联网快速发展的今天，Android 的开放性与兼容性越来越高，使 Android 移动应用的开发领域以及创意空间越来越大，开发人员可通过软硬件相结合的创新设计，打开移动应用新的大门。

本书以 U 酒保项目为基础，实现模块化的排列方式，最终以 Android 技术知识点为教学项目的形式展现给读者，使读者读完本书后，对 Android 应用开发有了系统的了解，并且具备了项目开发能力。

本书由八个项目模块组成，分别是深入 Android Studio、闪屏导航、登录注册、酒精检测、健康助手、打车代驾、无忧险、服务器部署与报错处理。循序渐进地讲述了 Android 项目开发的具体流程。通过本书的学习，读者可以更加熟练地使用 Android Studio 进行 Android 项目的开发，了解项目开发的流程与要点，设计出稳定高效的 App。

本书每个项目都按照 U 酒保开发流程进行讲解。都设有学习目标、学习路径、任务描述、任务技能、任务实施、任务总结、英语角以及任务习题。结构条理清晰、内容详细，任务实施可以将所学的理论知识充分的应用到实战中。

本书由归达伟、王磊任主编，熊祖涛、赵杰、高德梅、周青政、马高平任副主编，归达伟、熊祖涛、高德梅、马高平负责全面内容的规划，赵杰负责统稿、编排。具体分工如下：项目一至项目三由归达伟、王磊共同编写，归达伟负责全面规划；项目四至项目五由熊祖涛、赵杰共同编写，熊祖涛负责全面规划，项目六和项目七由高德梅、周青政共同编写，高德梅负责全面规划；项目八由马高平编写并负责全面规划。

本书理论内容简明、扼要；实例操作讲解细致，步骤清晰，实现了理实结合，操作步骤后有相对应的效果图，便于读者直观、清晰地看到操作效果，牢记书中的操作步骤。使读者在 Android 的学习过程中能够更加顺利，使自身的 Android 能力更上一层楼。

天津滨海迅腾科技集团有限公司

技术研发部

目　录

项目一　深入 Android Studio ……………………………………………………………… 1

　　学习目标 ……………………………………………………………………………… 1

　　学习路径 ……………………………………………………………………………… 1

　　任务描述 ……………………………………………………………………………… 1

　　任务技能 ……………………………………………………………………………… 2

　　　　技能点 1　项目分析 ……………………………………………………………… 2

　　　　技能点 2　导入 Eclipse 项目 …………………………………………………… 3

　　　　技能点 3　Android Studio 集成 Git 版本控制 ………………………………… 8

　　　　技能点 4　克隆 GitHub 项目到 Android Studio 上 ……………………………… 10

　　　　技能点 5　Android Studio 中 NDK 开发配置 …………………………………… 12

　　任务实施 ……………………………………………………………………………… 14

　　任务总结 ……………………………………………………………………………… 23

　　英语角 ………………………………………………………………………………… 23

　　任务习题 ……………………………………………………………………………… 23

项目二　闪屏导航 …………………………………………………………………………… 25

　　学习目标 ……………………………………………………………………………… 25

　　学习路径 ……………………………………………………………………………… 25

　　任务描述 ……………………………………………………………………………… 25

　　任务技能 ……………………………………………………………………………… 27

　　　　技能点 1　Android 原生动作 …………………………………………………… 27

　　　　技能点 2　TCP/IP …………………………………………………………………… 33

　　　　技能点 3　HttpURLConnection …………………………………………………… 34

　　　　技能点 4　PULL 解析 ……………………………………………………………… 39

　　任务实施 ……………………………………………………………………………… 42

　　任务总结 ……………………………………………………………………………… 54

　　英语角 ………………………………………………………………………………… 54

　　任务习题 ……………………………………………………………………………… 54

项目三　登录注册 …………………………………………………………………………… 56

　　学习目标 ……………………………………………………………………………… 56

　　学习路径 ……………………………………………………………………………… 56

　　任务描述 ……………………………………………………………………………… 56

任务技能 ································· 59

    技能点 1　SlidingMenu ················· 59

    技能点 2　ShareSDK ·················· 61

任务实施 ································· 63

任务总结 ································· 86

英语角 ·································· 86

任务习题 ································· 86

项目四　酒精检测 ························· 88

学习目标 ································· 88

学习路径 ································· 88

任务描述 ································· 88

任务技能 ································· 89

    技能点 1　蓝牙 ···················· 89

    技能点 2　进度条 ··················· 91

    技能点 3　复杂 JSON 解析 ··············· 95

任务实施 ································ 100

任务总结 ································ 115

英语角 ································· 115

任务习题 ································ 116

项目五　健康助手 ························ 117

学习目标 ································ 117

学习路径 ································ 117

任务描述 ································ 117

任务技能 ································ 118

    技能点 1　自定义组件 ················ 118

    技能点 2　自定义动画 ················ 123

    技能点 3　异步类 ·················· 128

任务实施 ································ 131

任务总结 ································ 145

英语角 ································· 145

任务习题 ································ 146

项目六　打车代驾 ························ 148

学习目标 ································ 148

学习路径 ································ 148

任务描述 ································ 148

任务技能 ································ 150

    技能点 1　Android 电话服务 ············· 150

技能点 2　Android MD5 加密 ·· 152

技能点 3　Stream 流 ·· 155

任务实施 ··· 160

任务总结 ··· 175

英语角 ·· 175

任务习题 ··· 176

项目七　无忧险 ··· 177

学习目标 ··· 177

学习路径 ··· 177

任务描述 ··· 177

任务技能 ··· 178

技能点 1　刷新加载 ·· 178

技能点 2　UI 更新 ·· 184

任务实施 ··· 187

任务总结 ··· 203

英语角 ·· 204

任务习题 ··· 204

项目八　服务器部署与报错处理 ·································· 206

学习目标 ··· 206

学习路径 ··· 206

任务描述 ··· 206

任务技能 ··· 207

技能点 1　服务器搭建 ·· 207

技能点 2　常见报错及解决方案 ···································· 210

任务实施 ··· 217

任务总结 ··· 221

英语角 ·· 222

任务习题 ··· 222

项目一　深入 Android Studio

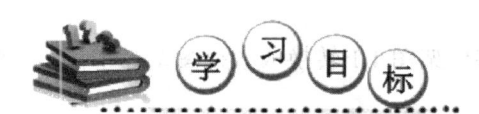

通过 U 酒保项目的学习，了解 Android Studio 内部的功能，创建 Android 项目，掌握硬件通信原理，具有独立创建和编写项目的能力。在任务实现过程中：

- 学习 Eclipse 项目导入 Android Studio 的步骤。
- 了解 Android Studio 项目上传到 GitHub 上面的方法。
- 掌握 Android Studio 从 GitHub 上克隆项目的方法。

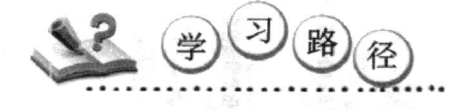

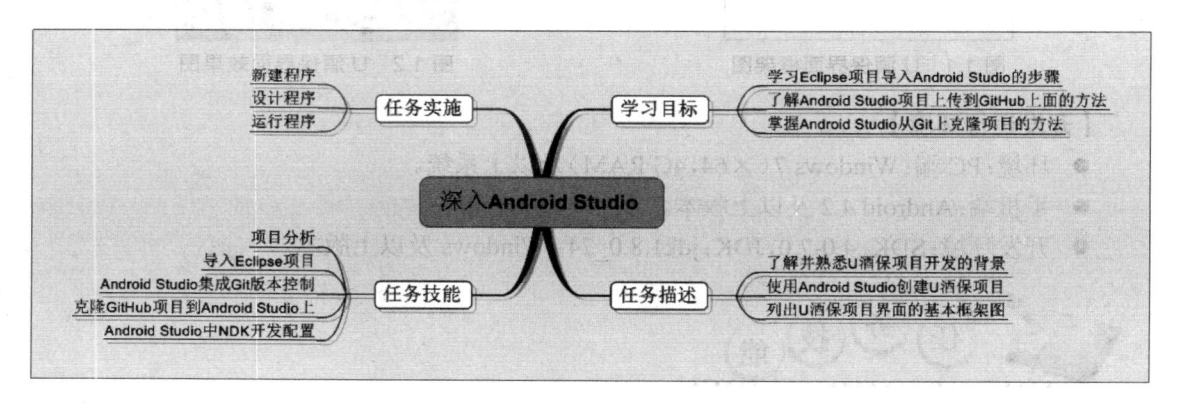

【情境导入】

在日常生活中，交通事故所引起的安全问题已经成为人身安全的最大威胁。世界上每年有几十万人在车祸中失去生命。而造成这些交通事故的因素有人为、车辆、路况、其他外界环境与管理等方面，其中酒驾是发生悲剧的重要原因之一，为减少悲剧的发生，研发人员设计了一款基于 Android 平台的便携式酒精检测系统——U 酒保，减少酒驾的同时也为安全提供保障。该项目主要讲解在 Android 项目的开发过程中所需软件的安装和环境配置,并实现项目创建。

【功能描述】

本模块使用 Android Studio 创建 U 酒保项目:

● 创建 Android 项目。
● 在虚拟机上运行项目。
● 进行项目调试。

【基本框架】

基本框架如图 1.1 所示。通过本模块的学习,能将框架图 1.1 转换成效果图 1.2 所示。

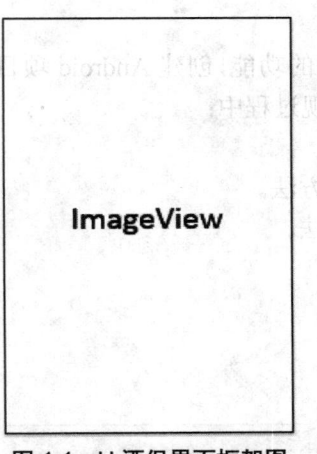

图 1.1 U 酒保界面框架图

图 1.2 U 酒保界面效果图

【开发运行环境】

● 环境:PC 端:Windows 7(×64,4G RAM)及以上系统。
● 手机端:Android 4.2 及以上版本。
● 开发环境:SDK:4.0-7.0,JDK:jdk1.8.0_74 –Windows 及以上版本。

技能点 1 项目分析

1 U 酒保项目背景

U 酒保是一款便携式酒精检测系统,开发的目的是为了减少酒驾,同时为安全出行提供保障。本系统基于 Android 开发环境,运用 MVC 开发模式进行编写,项目中使用了扁平化的 UI

设计使用户在体验过程中获得更高的舒适感,数据间使用蓝牙模块进行数据传输,通道使用 Socket。该项目主要分为三大模块,分别是:用户初次进入时的欢迎导航界面,用户登录时的登录界面以及有核心功能的主界面。主界面中主要有酒精检测、健康助手、打车代驾、无忧险四个功能模块如图 1.3 所示。

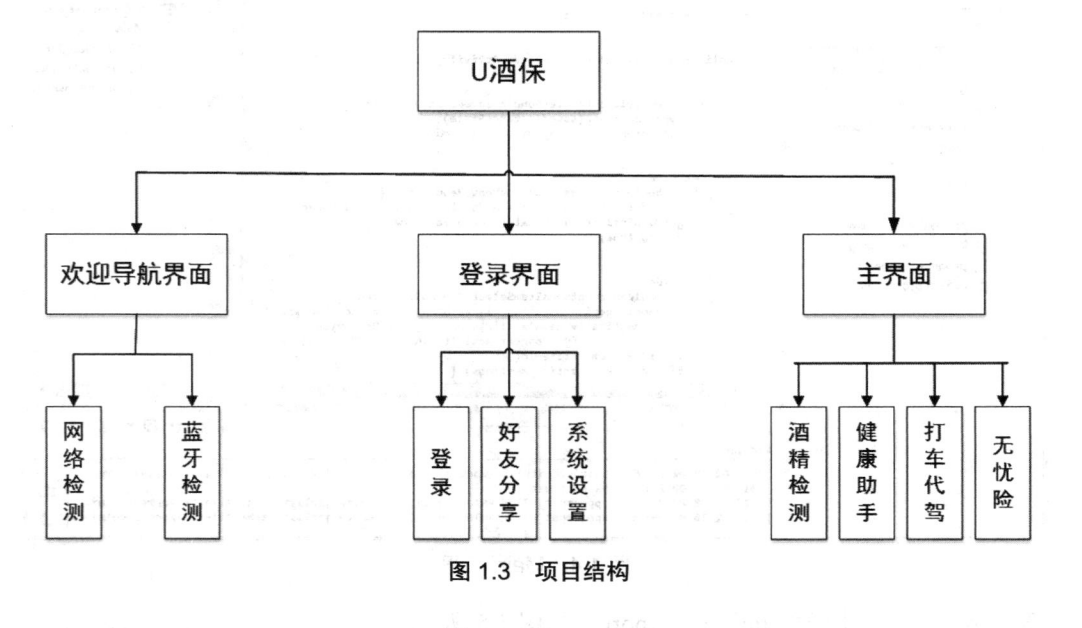

图 1.3 项目结构

2 酒精检测仪器介绍

硬件设备采用 MQ-3 气体传感器与 STM32F103R 开发板,数据间使用蓝牙模块进行数据传输,通道使用 Socket。通过蓝牙连接 MQ-3 气体传感器,将检测参数接入到酒精浓度检测模块中,通过模拟电压信号放大判断酒精浓度,将采集到的模拟电压信号通过单片机控制经 A/D 转换,得到数字电压信号,用于显示浓度的数码管显示模块,通过电压到浓度的线性转换和最终浓度值的数码管显示。

技能点 2 导入 Eclipse 项目

将 Eclipse 项目导入 Android Studio 中步骤如下。

第一步:在 Eclipse 中新建项目命名为"Main",如图 1.4 所示。

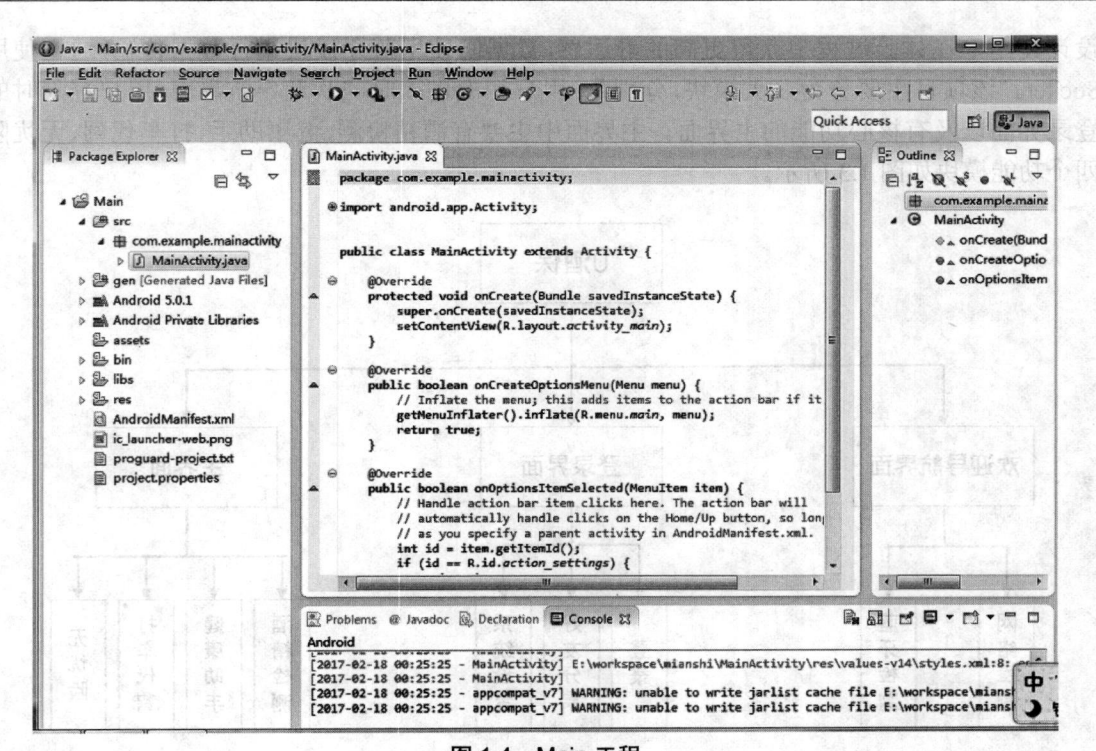

图 1.4 Main 工程

第二步：选择左上角"File"→"Export"，如图 1.5 所示。

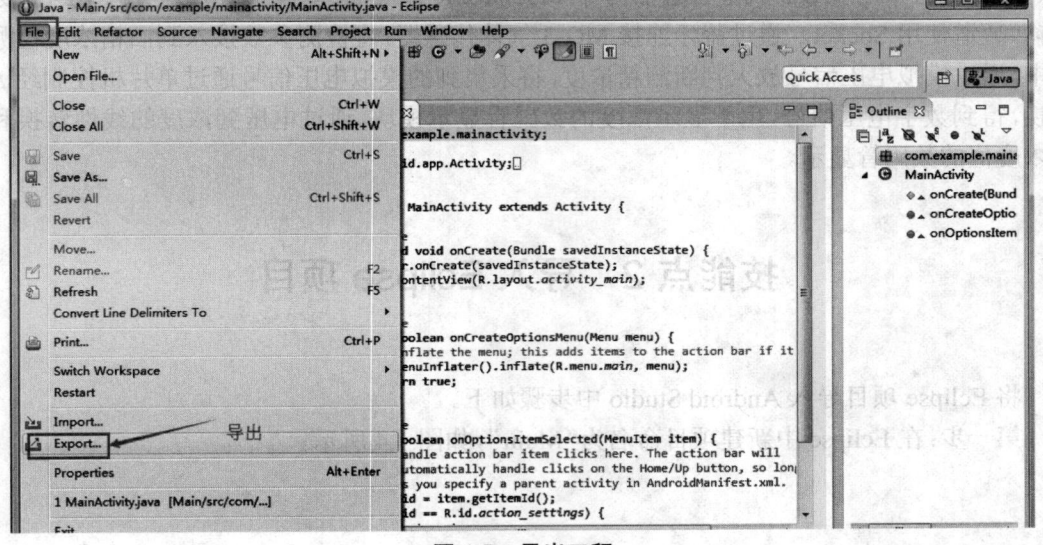

图 1.5 导出工程

第三步：选择 Android 下的"Generate Gradle build files"，点击"Next"，如图 1.6 所示，出现如图 1.7 所示界面，点击"Next"。

项目一 深入 Android Studio   5

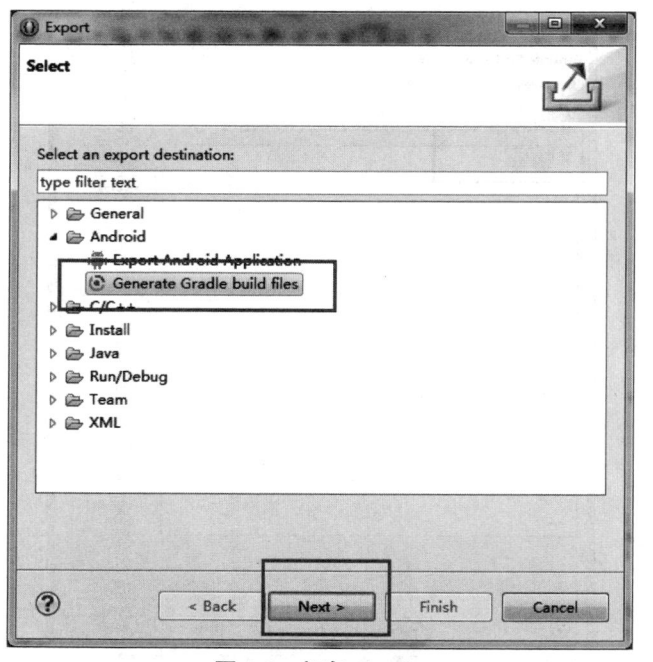

图 1.6 打包 Gradle

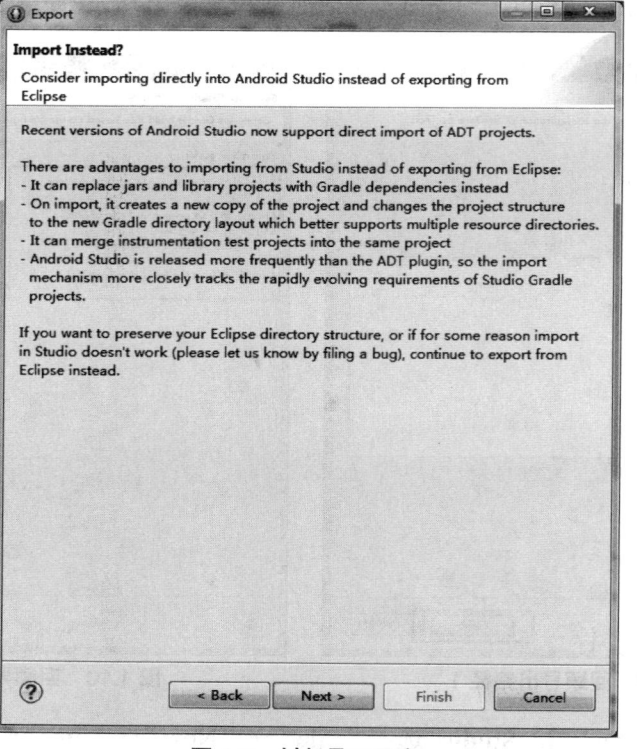

图 1.7 判断是否导出

第四步：选择对应工程后，点击"Next"，如图 1.8 所示。

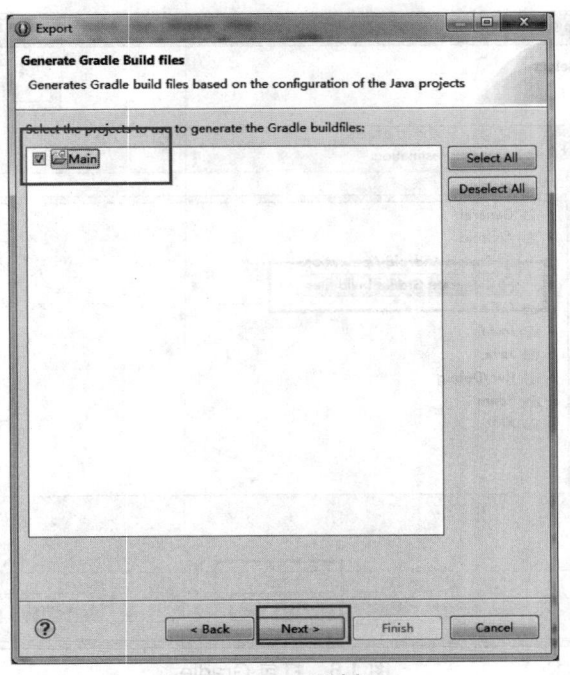

图 1.8　工程选择

第五步：记住导出路径，点击"Finish"，如图 1.9 和图 1.10 所示。

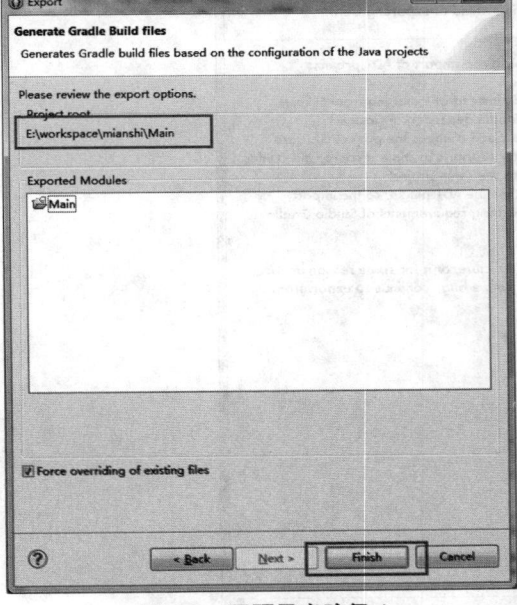

图 1.9　回顾导出路径 1

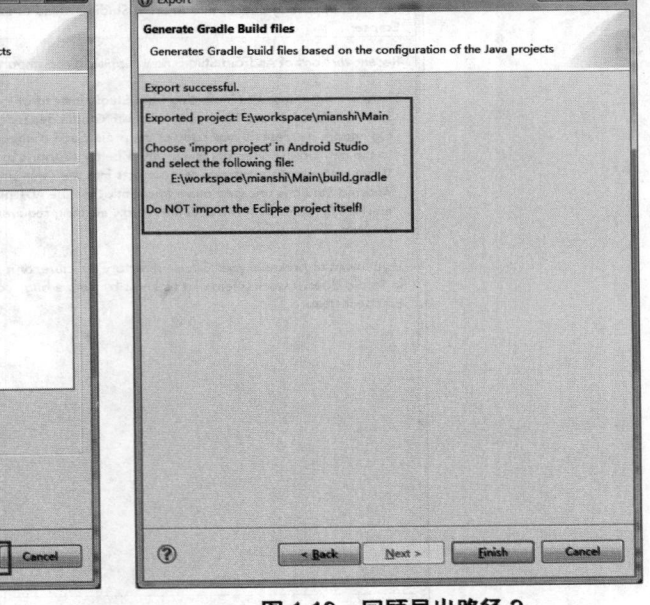

图 1.10　回顾导出路径 2

第六步：打开 Android Studio 选择"File"→"New"→"Import Project"，如图 1.11 所示。

项目一　深入 Android Studio　　　　7

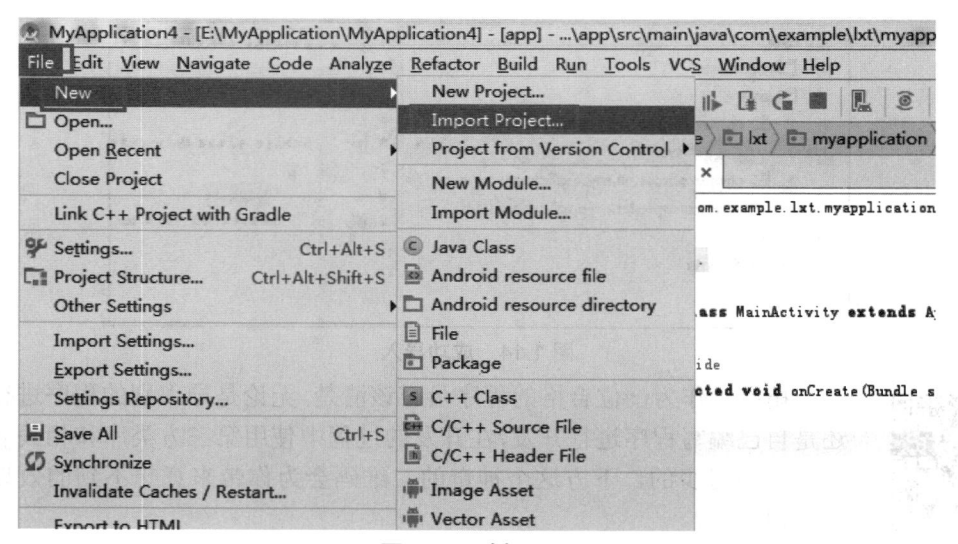

图 1.11　选择导入

第七步：根据工程路径选择要导入的工程，点击"OK"，如图 1.12 所示。

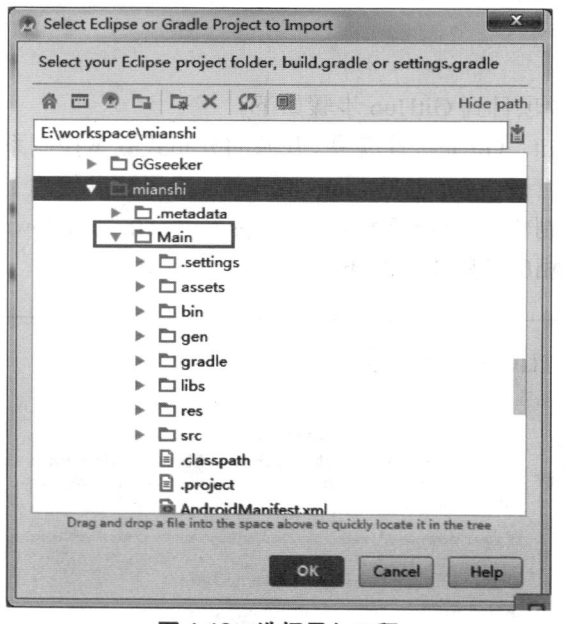

图 1.12　选择导入工程

第八步：项目正在导入，如图 1.13 所示。成功导入，如图 1.14 所示。

图 1.13　正在导入

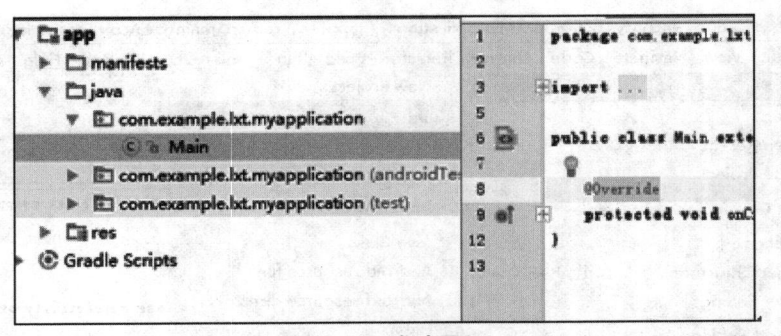

图 1.14　成功导入

拓展　作为一位合格的程序员应该清楚,无论是导入别的程序进行开发,还是自己编写程序进行开发,在开发的过程中使用第三方类库和相关 jar 包是不可必不可少的。下方这个神奇的二维码会为你带来意想不到的效果,还不赶快扫一下?

技能点 3　Android Studio 集成 Git 版本控制

上传 Android Studio 项目到 GitHub 步骤如下。

第一步:下载安装 Git。Git 是一个免费、开源的分布式版本控制系统,可以有效、高速的处理从小到大的项目版本管理。Git 也是 Linus Torvalds 为了帮助管理 Linux 内核开发而开发的一个开放源码的版本控制软件。Git 下载地址:https://git-scm.com/download/win。如图 1.15,根据自己的电脑系统选择相应的版本进行下载。

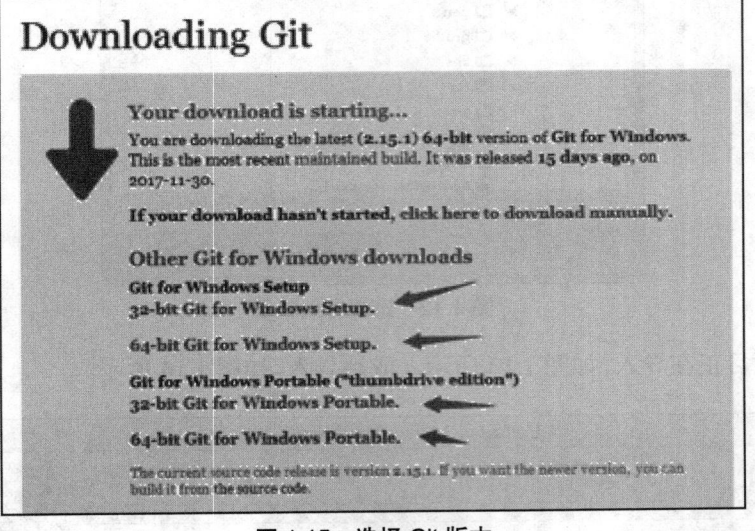

图 1.15　选择 Git 版本

下载以后进行安装,Git 的安装比较简单,这里就不详细介绍了。

项目一　深入 Android Studio　　　　　　　　　　　　9

第二步：集成 Git 到 Android Studio 上，如图 1.16 所示。

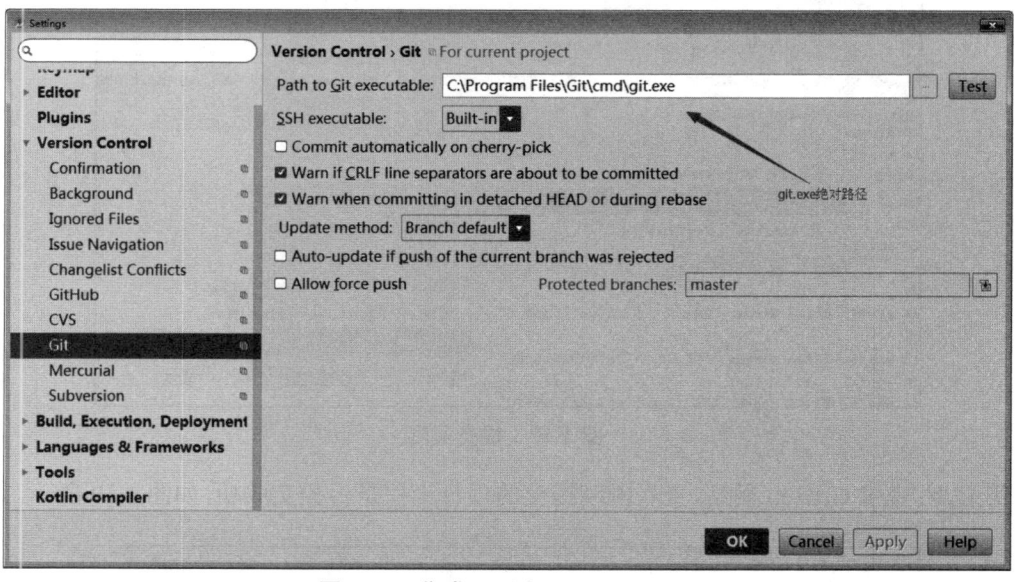

图 1.16　集成 Git 到 Android Studio

第三步：配置 Android Studio 中的 GitHub 账户，如图 1.17 所示。可以通过"Test"测试账号是否可用，如果没有 GitHub 的账号需要到其官网进行注册。

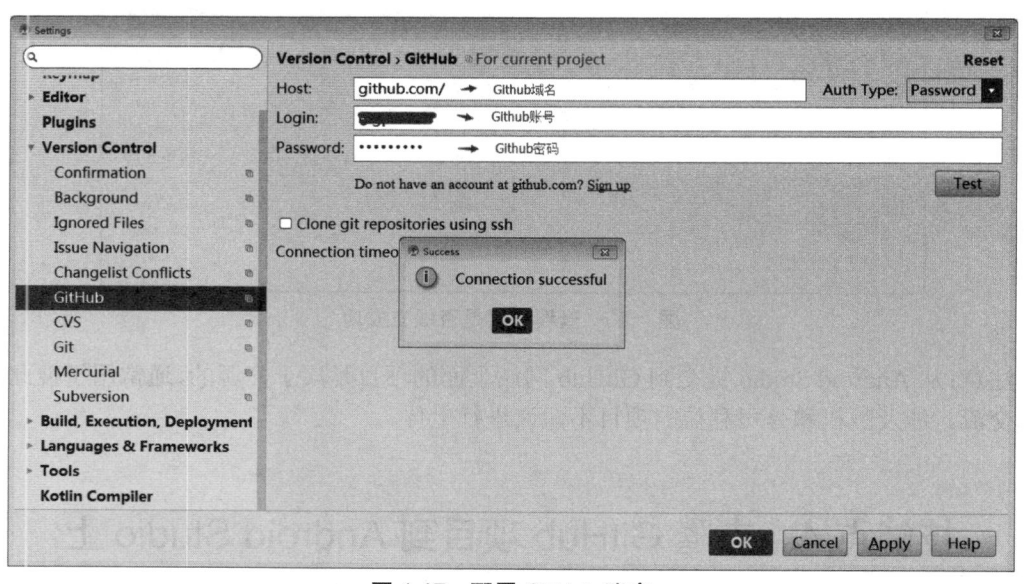

图 1.17　配置 GitHub 账户

第四步：把 Android Studio 上面的项目提交到 GitHub 远程仓库，如图 1.18 所示。

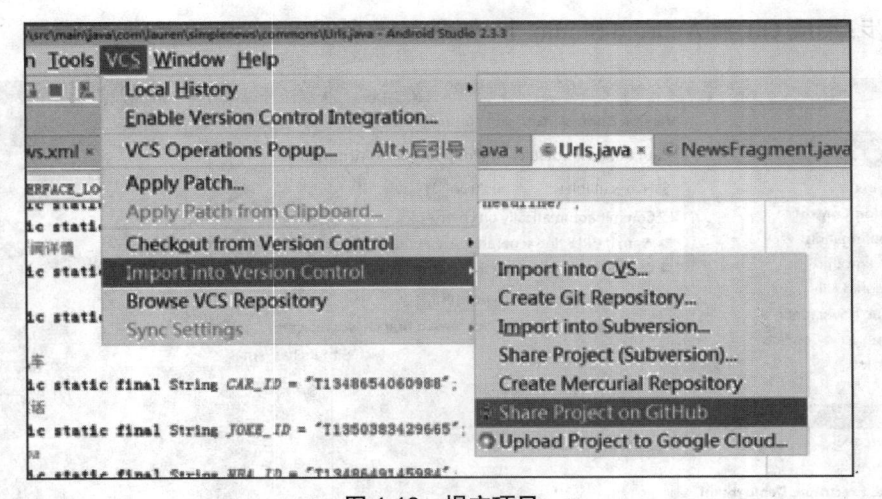

图 1.18　提交项目

第五步：登录 GitHub 网站，查看刚刚提交的项目工程是否提交成功，如图 1.19 所示。

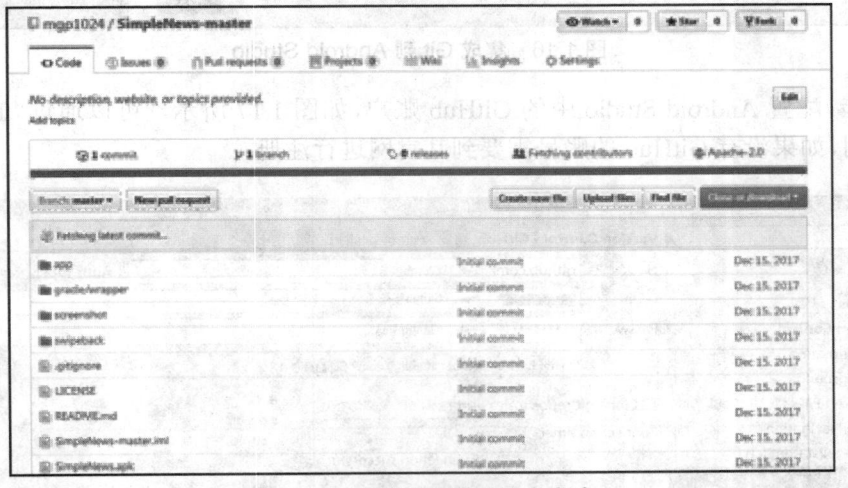

图 1.19　查看项目是否提交成功

注意：从 Android Studio 提交到 GitHub 网站上面的项目是属于开源的，通常用于彼此的学习和交流。涉及版权和公司利益的项目不建议进行上传。

技能点 4　克隆 GitHub 项目到 Android Studio 上

将 GitHub 上的项目克隆到 Android Studio 上的步骤如下。

第一步：在 Android Studio 中进行操作"VCS → Checkout from Version Control → GitHub"，如图 1.20 和图 1.21 所示。指定要克隆项目的 GitHub 地址。SimpleNews-master 项目地址：https://github.com/mgp1024/SimpleNews-master。

项目一 深入 Android Studio 11

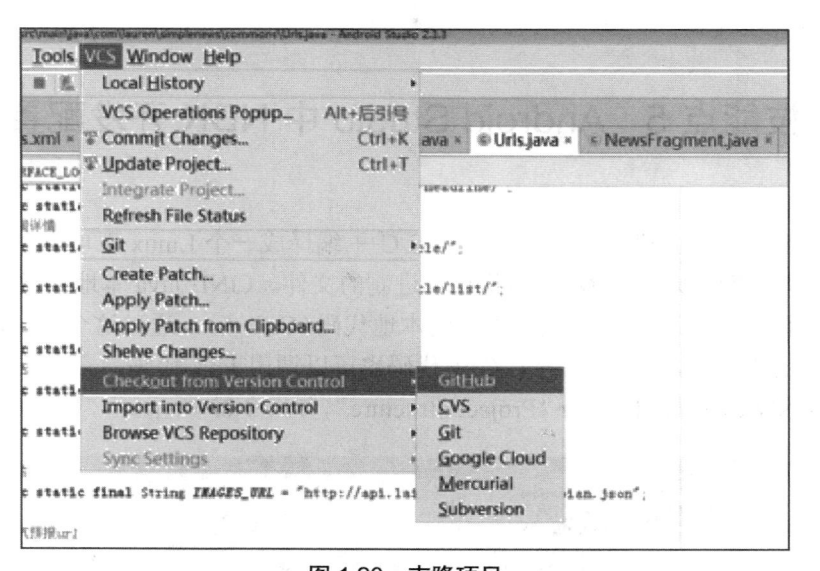

图 1.20 克隆项目

图 1.21 详细地址

第二步：克隆完毕后，查看 SimpleNews-master 项目目录结构，如图 1.22 所示。

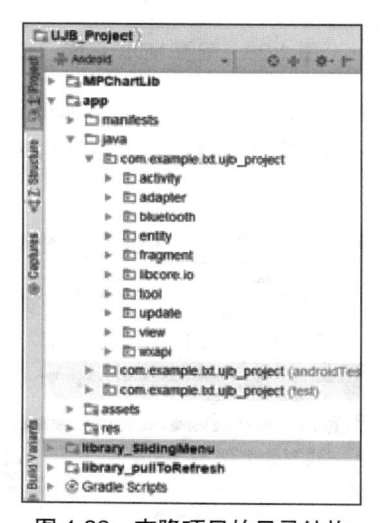

图 1.22 克隆项目的目录结构

技能点 5　Android Studio 中 NDK 开发配置

NDK：Android 本地开发工具集，可以把 C/C++ 编译成一个 Linux 下可以执行的二进制文件，Java 代码里面就可以通过 JNI 调用执行二进制的文件。（JNI：Java 本地开发接口，JNI 是一个协议这个协议用来沟通 Java 代码和外部的本地代码（C/C++））。通过这个协议，Java 代码就可以调用外部的 C/C++，代码外部的 C/C++ 代码也可以调用 Java 代码。）

（1）配置 NDK 环境："File"→"Project Structure"，如图 1.23 所示。

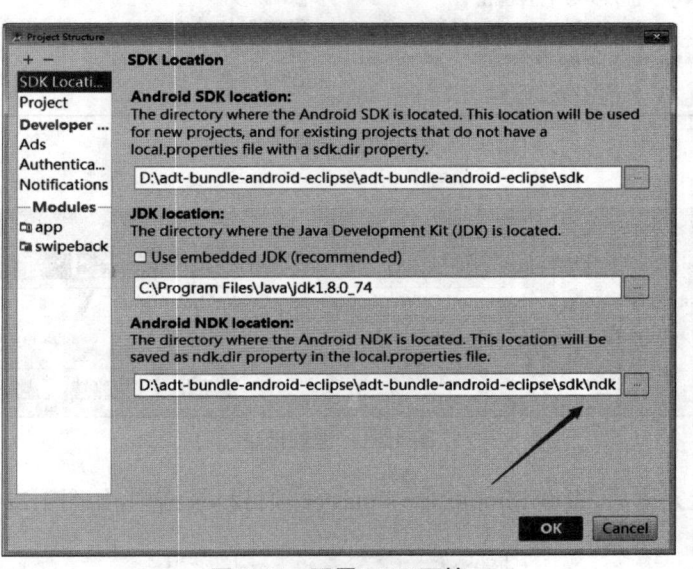

图 1.23　配置 NDK 环境

（2）打开下载 SDK 界面安装 CMake："Tools"→"Android"→"SDK Manager"，如图 1.24 所示。其中 CMake 是构建 C/C++ 代码的工具，根据谷歌官方描述 CMake 开发 NDK 只支持 64 位系统，32 位系统是无法使用的。

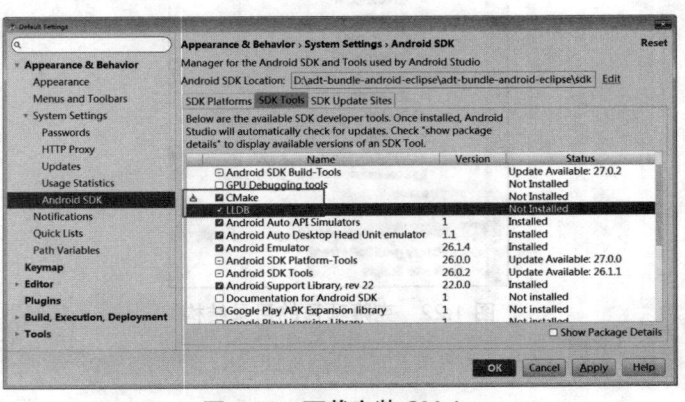

图 1.24　下载安装 CMake

CMake：CMake 是一个跨平台的安装（编译）工具，可以用简单的语句来描述所有平台的安装（编译过程）。

（3）下载完成之后创建支持 C/C++ 开发的 Android 项目，如图 1.25 所示。注意要选中箭头所指部分。

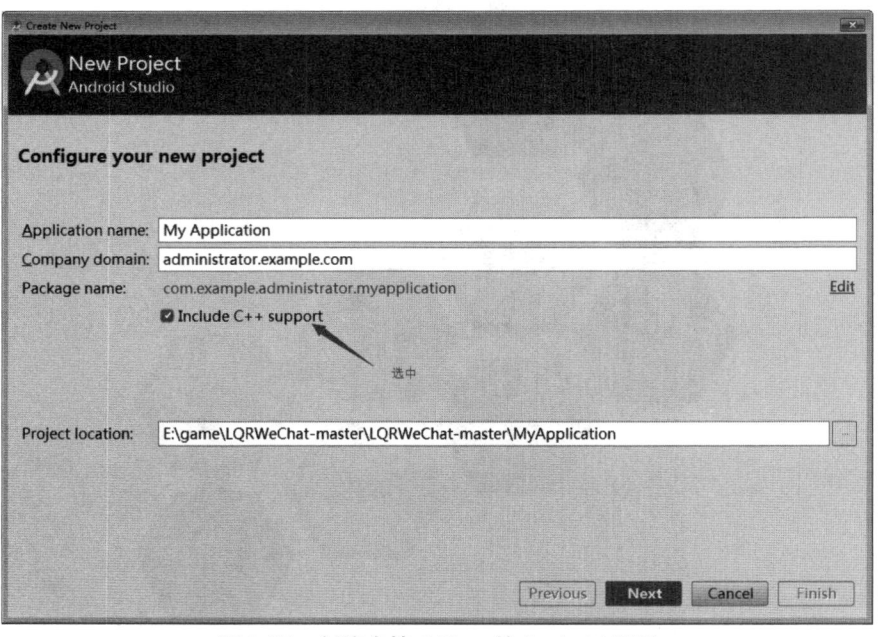

图 1.25　创建支持 C/C++ 的 Android 项目

在创建项目之后，打开项目的目录，发现有如图 1.26 画框部分的文件则证明支持 C/C++ 开发的 Android 工程创建成功了。

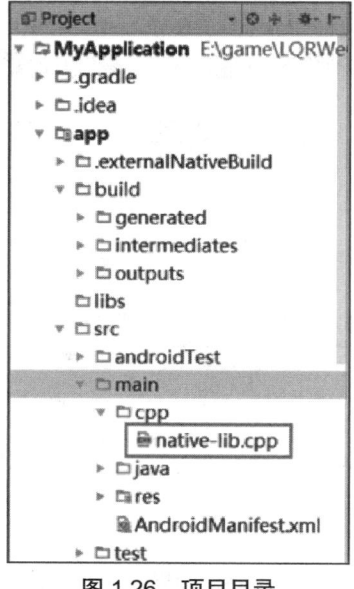

图 1.26　项目目录

14　　　　　　　　　　　　　　Android 模块化项目实战

通过如下步骤实现在 Android Studio 中创建第一个 Android 项目。

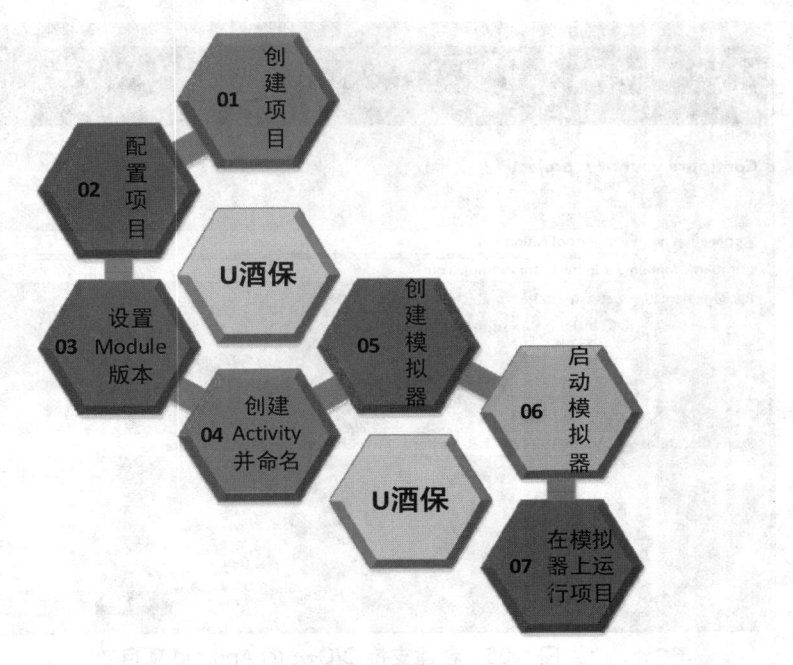

具体步骤如下所示。

第一步：打开 Android Studio，新建第一个项目。如果未打开项目，在 Welcome to Android Studio 窗口中，点击"Start a new Android Studio project"新建项目，如图 1.27 所示。如果已打开项目请选择"File"→"New Project"，如图 1.28 所示。

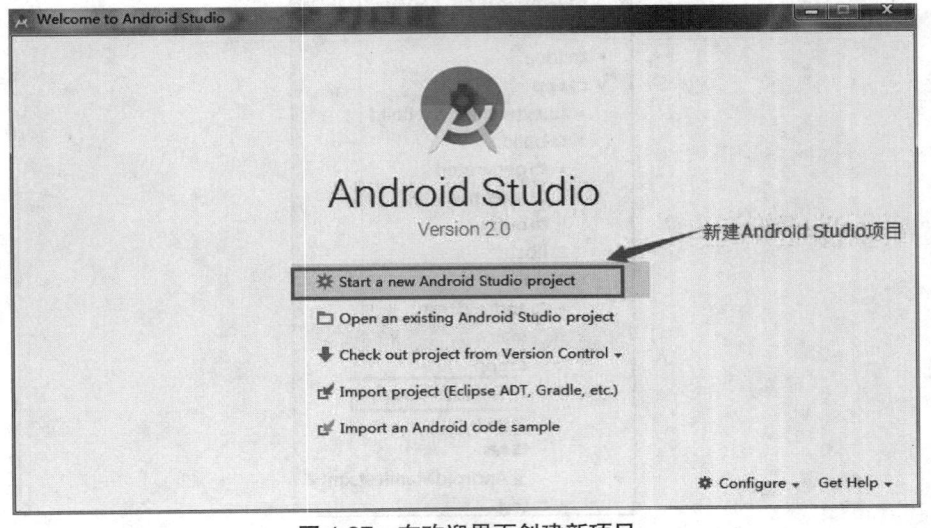

图 1.27　在欢迎界面创建新项目

项目一　深入 Android Studio

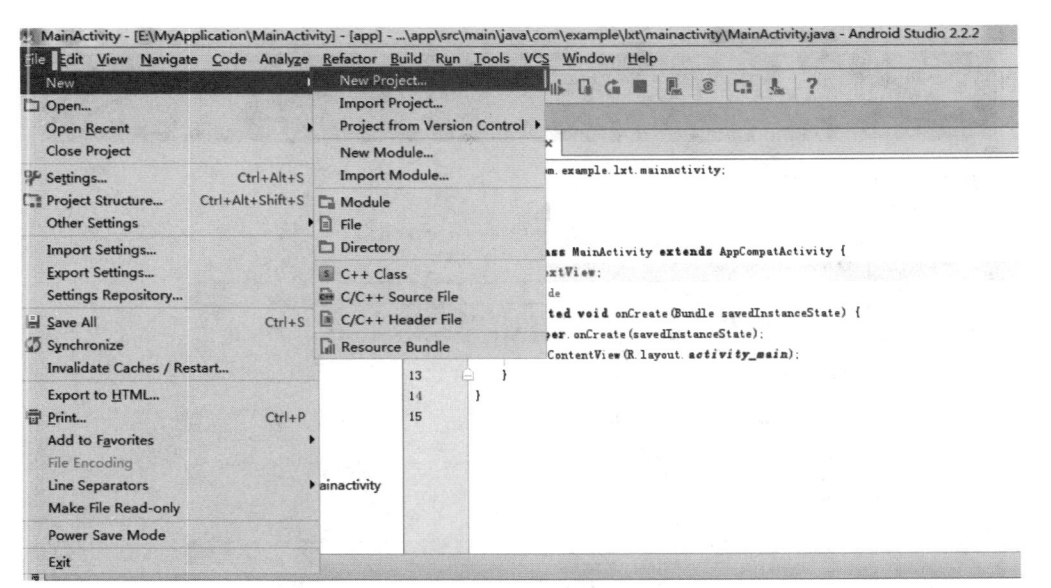

图 1.28　在工具中创建新项目

第二步：在创建工程页面中配置项目，添加项目名称、域名、以及存放路径，设置完成后点击"Next"，如图 1.29 所示。

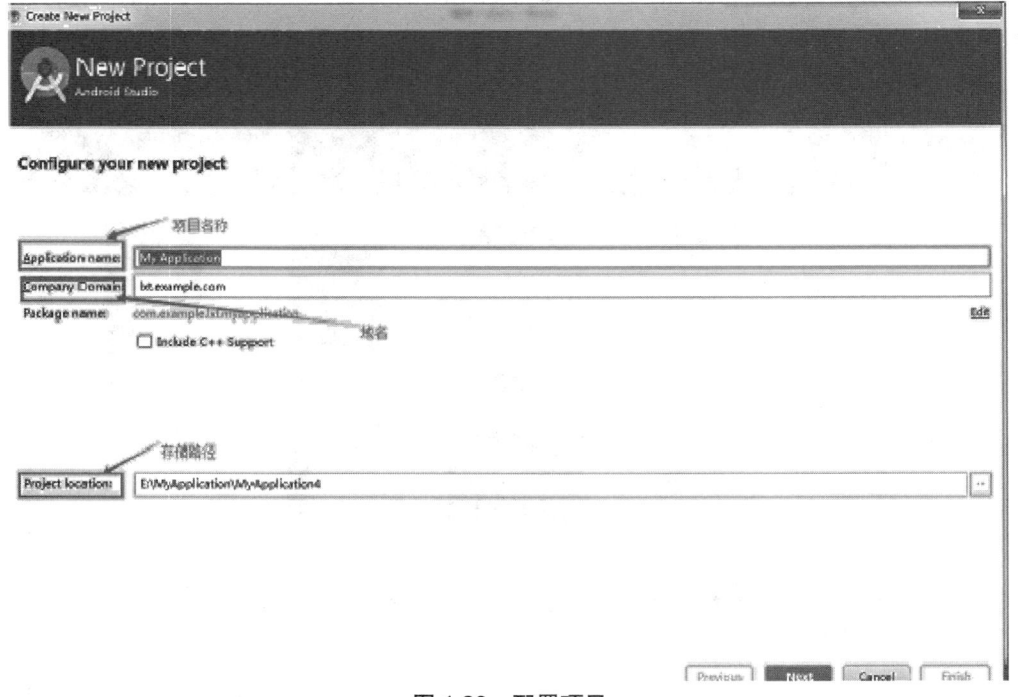

图 1.29　配置项目

第三步：在 Phone and Tablet（手机和平板项目）下的 Minimun SDK 中设置 Module 支持的 Android 兼容最低版本，根据不同的用户可选择不同的版本，可点击"Help me choose"查看当前 Android 版本分布情况，如图 1.30 所示，设置完成后点击"Next"。

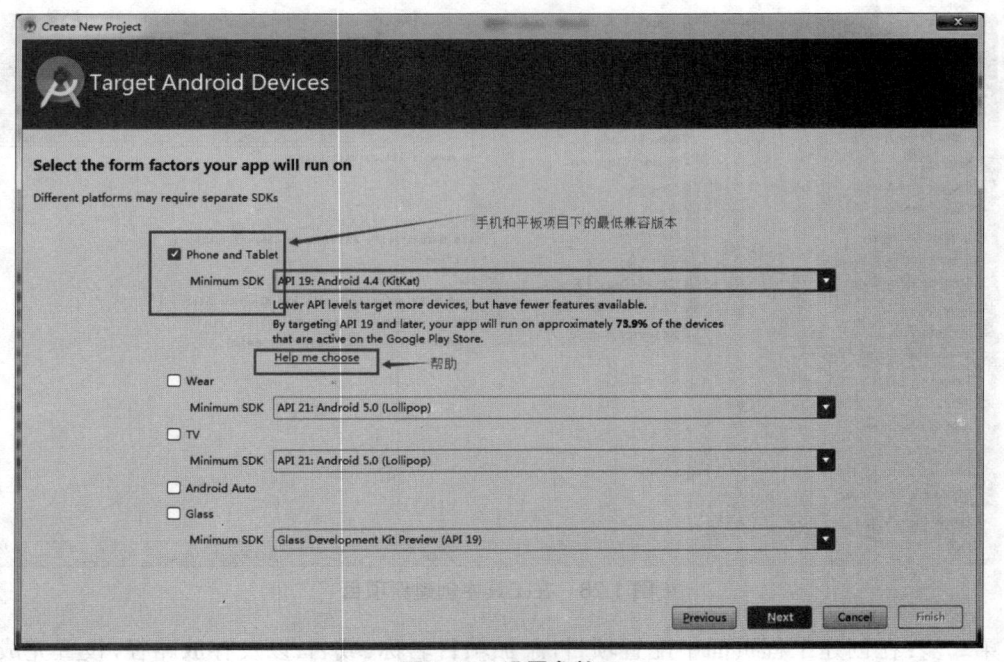

图 1.30　设置参数

第四步：选择是否创建 Activity 以及创建 Activity 的类型。默认选择 Empty（空的）Activity，如图 1.31 所示，选择完成后点击"Next"。

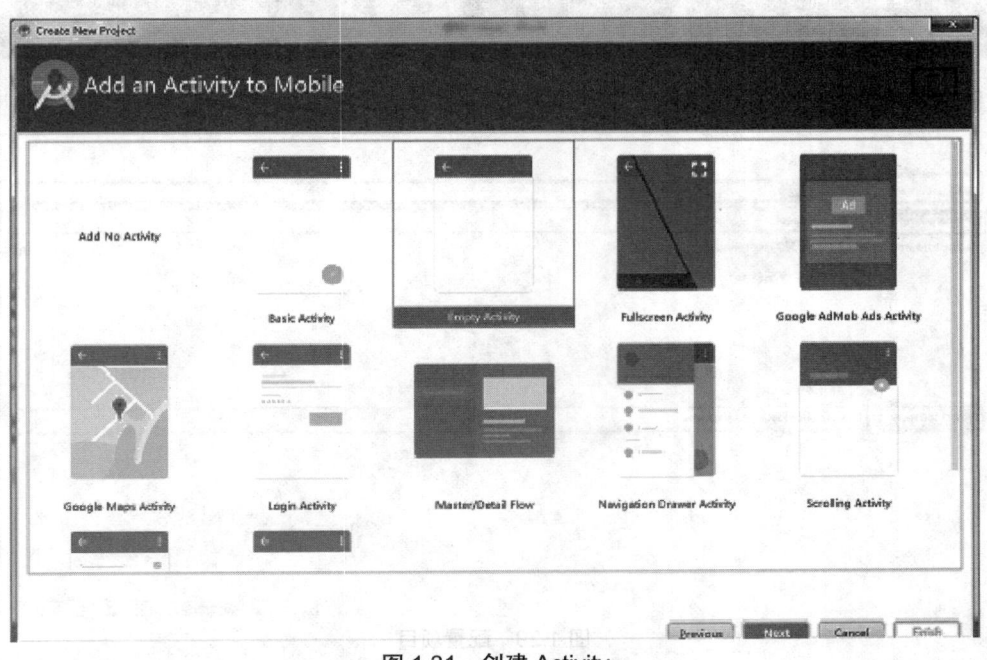

图 1.31　创建 Activity

第五步：为 Empty Activity 添加名称，默认为 MainActivity，如图 1.32 所示。添加完成后点击"Finish"后出现进度条，如图 1.33 所示。

项目一 深入 Android Studio 17

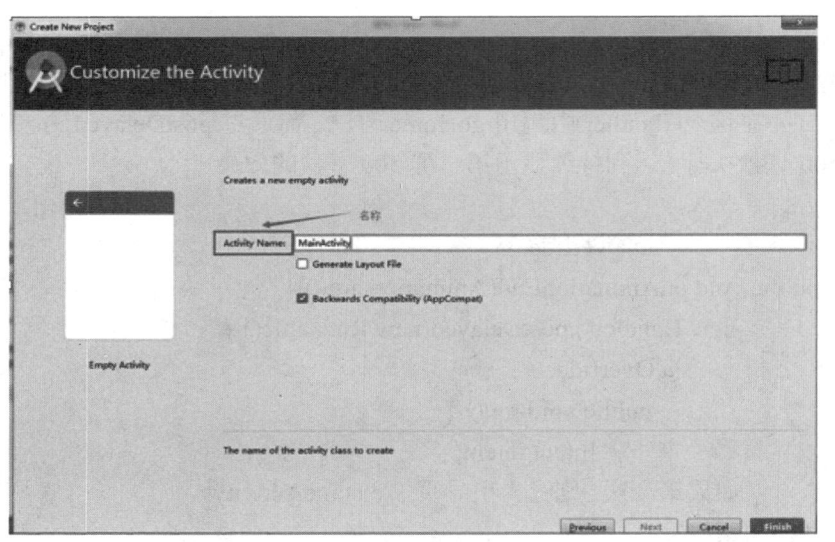

图 1.32 添加 Activity 名

图 1.33 创建编译项目

第六步：项目创建完成后设置界面布局并添加相关代码实现初次进入程序时的导航效果，如代码 CORE0101 所示。

代码 CORE0101 闪屏界面

```
private void FirstIntent() {
/* 如果网络可用则判断是否第一次进入，如果是第一次则进入欢迎界面 */
    first = shared.getBoolean("First", true);
/* 设置动画效果是 alpha, 在 anim 目录下的 alpha.xml 文件中定义动画效果 */
    animation = AnimationUtils.loadAnimation(this, R.anim.alpha);
        // 给 view 设置动画效果
    view.startAnimation(animation);
    animation.setAnimationListener(new AnimationListener() {
        @Override
        public void onAnimationStart(Animation arg0) {

        }
        @Override
        public void onAnimationRepeat(Animation arg0) {

        }
```

```
   /*
   这里监听动画结束的动作，在动画结束的时候开启一个线程，这个线程中绑定一
个 Handler, 并在这个 Handler 中调用 goHome 方法，而通过 postDelayed 方法使这个方
法延迟 500 毫秒执行，达到持续显示第一屏 500 毫秒的效果
   */
                       @Override
          public void onAnimationEnd(Animation arg0) {
                new Handler().postDelayed(new Runnable() {
                    @Override
                    public void run() {
                           Intent intent;
            // 如果第一次，则进入引导页 WelcomeActivity
          if (first) {
          intent = new Intent(WelcomActivity.this,GuideActivity.class);
                } else {
          intent = new Intent(WelcomActivity.this,MainActivity.class);
                              }
                           startActivity(intent);
          // 设置 Activity 的切换效果
          OverridePendingTransition(R.anim.in_from_right,R.anim.out_to_left);
                       WelcomActivity.this.finish();
                    }
                }, TIME);
          }});}
```

编写完代码后进入 Android Studio 主页面，点击机器人图标创建模拟器，如图 1.34 所示。

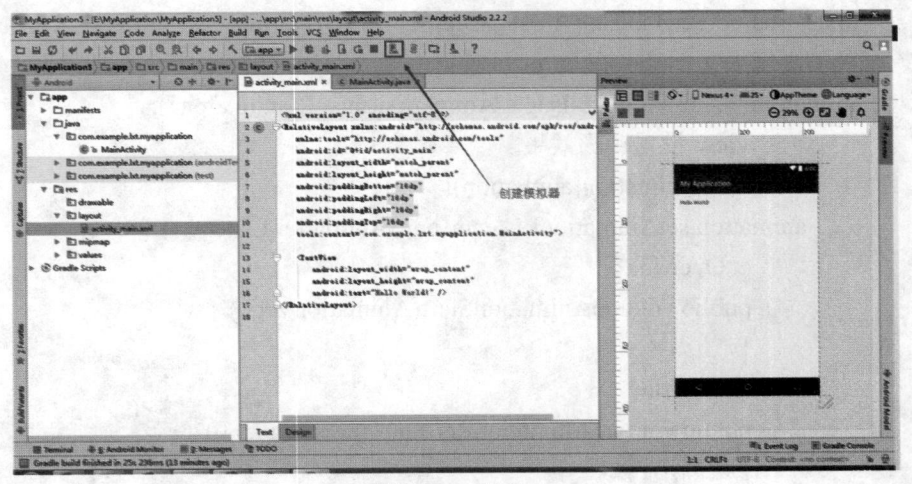

图 1.34 主页面

项目一　深入 Android Studio　　　　　　　　　　　　　　　　　　　　　19

　　第七步：点击"Create Virtual Device..."按钮，如图 1.35 所示。进入模拟器规格选择界面，如图 1.36 所示，选择模拟器的规格，默认为 Phone，其中选择 Nexus One，点击"Next"。进入下一界面如图 1.37 所示，点击"Finish"，至此模拟器创建完成。

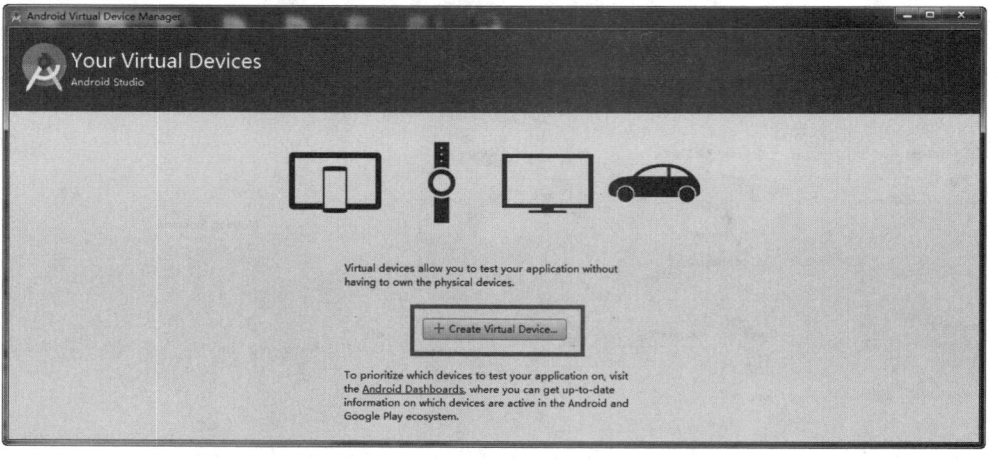

图 1.35　创建模拟器

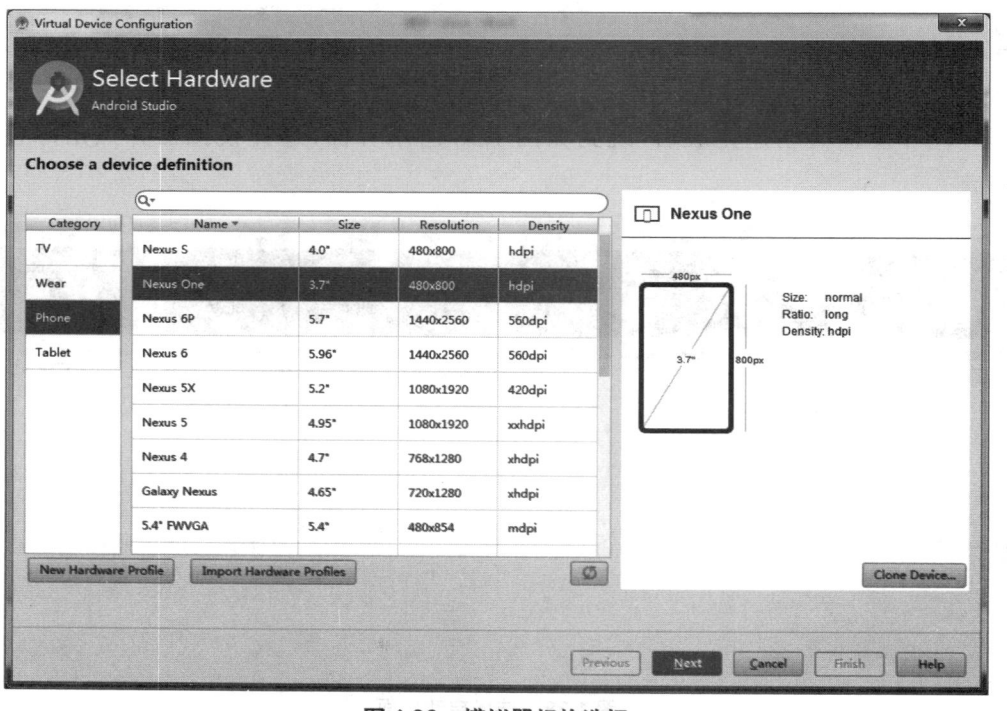

图 1.36　模拟器规格选择

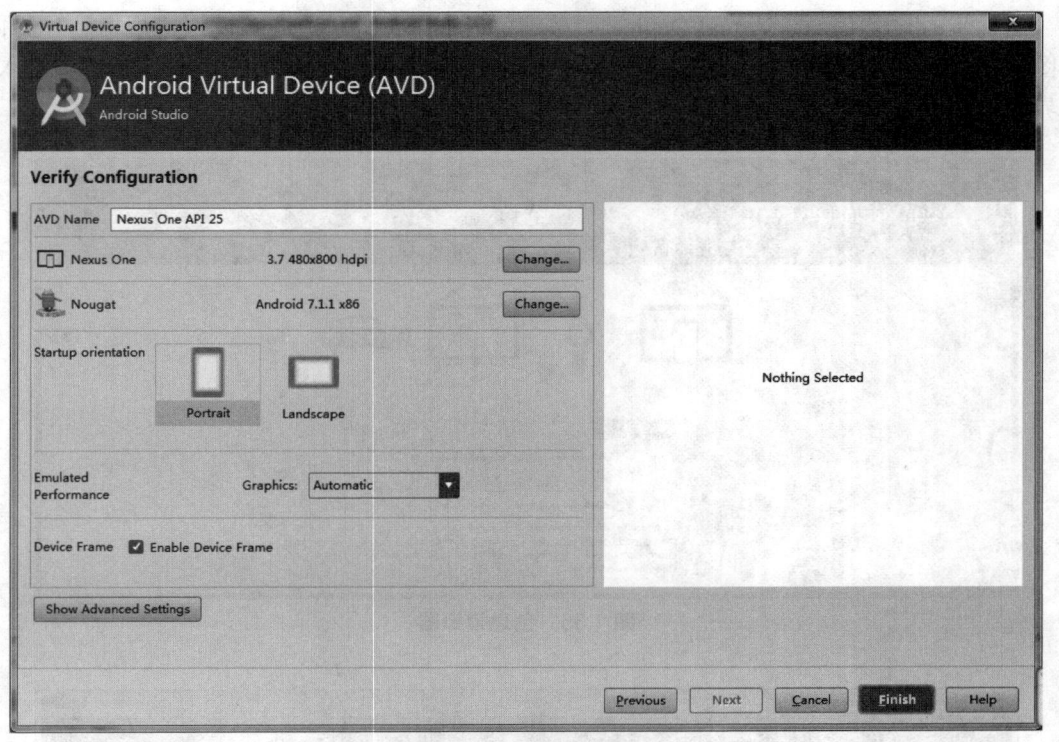

图 1.37　模拟器构造

第八步：模拟器创建完成后，跳到如图 1.38 所示界面，选择模拟器并点击绿色箭头启动。

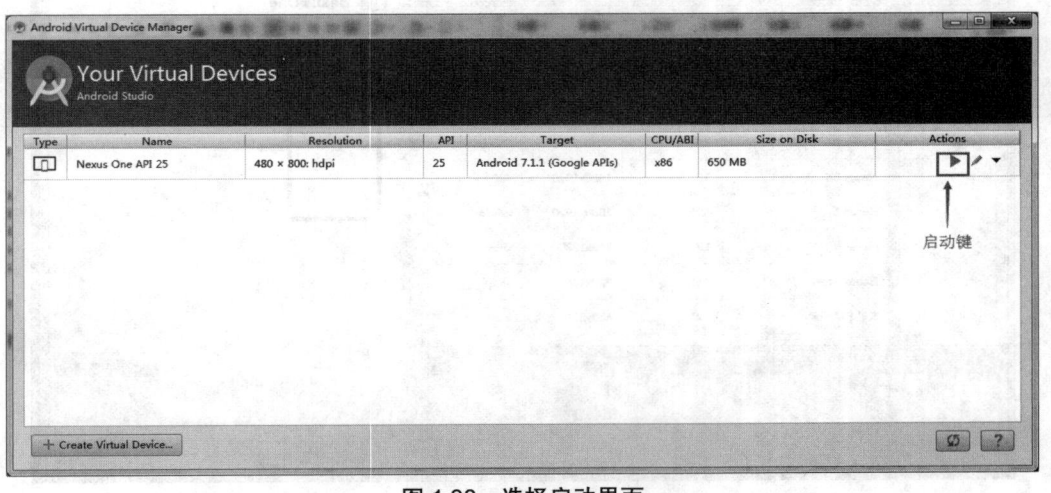

图 1.38　选择启动界面

第九步：模拟器界面如图 1.39 所示。

项目一 深入 Android Studio 21

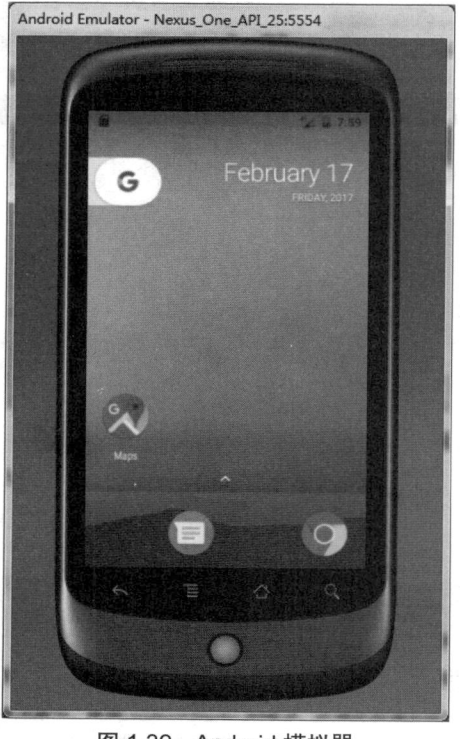

图 1.39 Android 模拟器

第十步：进入 Android Studio 中点击绿色箭头运行程序，如图 1.40 所示，选择所使用的模拟器，点击"OK"，在模拟器上运行，如图 1.41 所示。

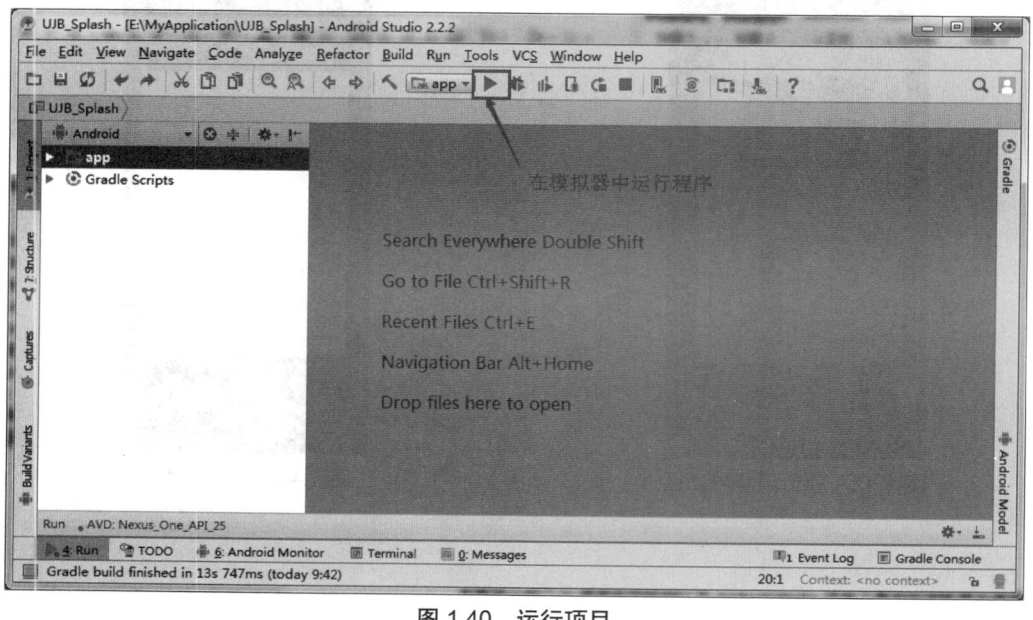

图 1.40 运行项目

22 　　　　　　　　　　　　Android 模块化项目实战

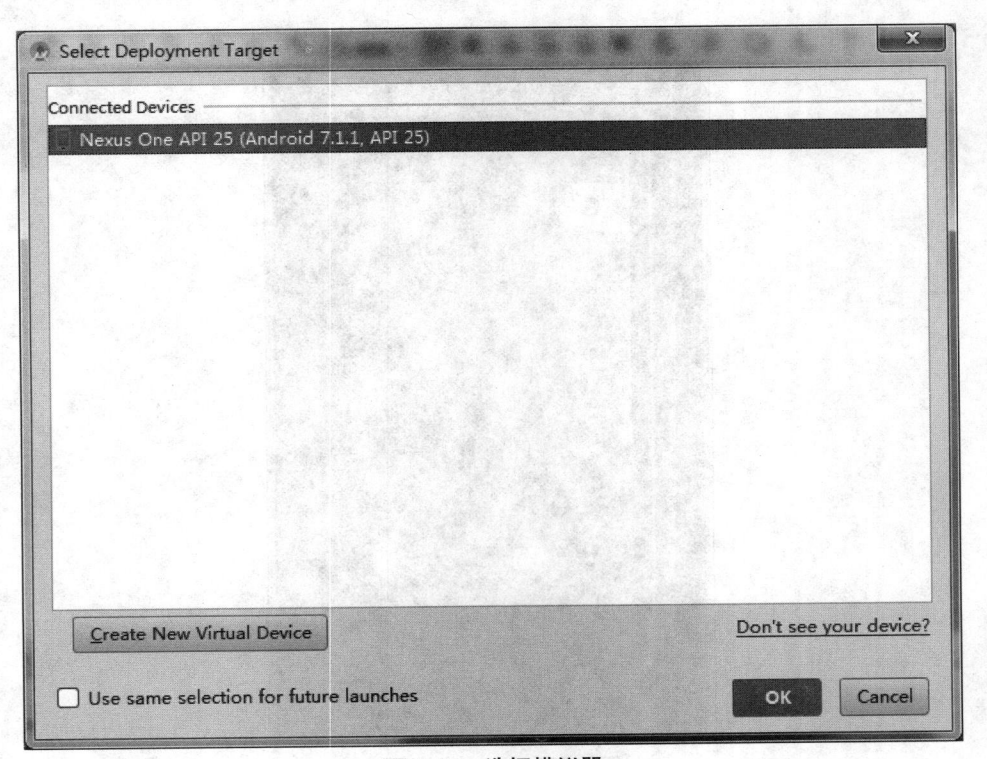

图 1.41　选择模拟器

第十一步：在模拟器中显示运行效果，如图 1.42 和图 1.43 所示。

图 1.42　项目闪屏界面图

图 1.43　项目主界面

项目一 深入 Android Studio

本项目介绍了 U 酒保项目结构以及 Android Studio 开发工具的使用方法,重点讲解如何使用 Android Studio 开发工具。通过对本项目的学习可以清楚的了解 Android 开发的基本概念,掌握 Android Studio 的特点、使用技巧和程序运行的基本流程,提高对 Android 项目的认知度以及 Android Studio 使用熟练度。

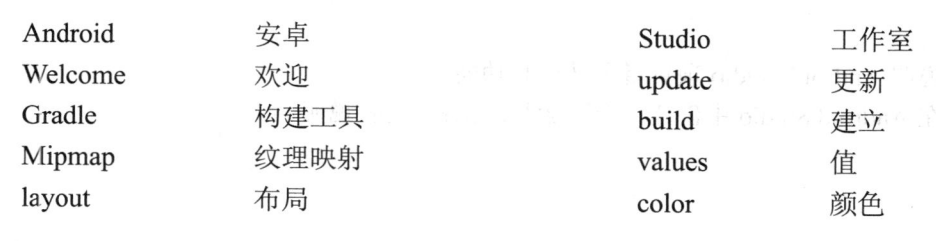

Android	安卓	Studio	工作室
Welcome	欢迎	update	更新
Gradle	构建工具	build	建立
Mipmap	纹理映射	values	值
layout	布局	color	颜色

一、选择题

1. Android Studio 编辑器不仅吸收了( )的优点,还自带界面实时预览,并且界面显示非常清晰,便于修改。

A. Andriod

B. Eclipse

C. MyEclipse+ADT

D. Eclipse+ADT

2. Android Studio 项目中 res 目录中( )来存储布局文件。

A. drawable B. layout C. mipmap D. build

3. 在项目开发过程中,快捷键的使用能够实现程序快速准确的编写,提高代码编写效率。自动补全代码的快捷键为( )。

A. Alt+ 回车

B. Alt+/

C. Ctrl+Alt+L

D. Ctrl+Shift+Space

4. 下面选项中( )是创建一个新的工程。

A. Start a new Android Studio project

B. Import project

C. Open an existing Android Studio project

D. Import an Android code sample

5. 在 Phone and Tablet(手机和平板项目)下的( )中设置 Module 支持的 Android 兼容最低版本,根据不同的用户可选择不同的版本。

A. Minimun SDK B. Wear C. TV D. Class

二、填空题

1. 谷歌在 2013 年 5 月 16 日 I/O 大会上推出新的 Android 开发环境——Android Studio,并对开发者控制台进行了改进,增加了五个新功能,分别为 ＿＿＿＿＿、＿＿＿＿＿、＿＿＿＿＿、＿＿＿＿＿、试用版测试和阶段性展示。

2. Android Studio 在安装之前应该先 ＿＿＿＿＿＿＿＿＿＿＿＿＿＿＿＿＿＿＿＿＿＿＿＿＿＿＿。

3. Android Studio 需要配置环境变量分别为 ＿＿＿＿＿、＿＿＿＿＿、＿＿＿＿＿。

4. Android Studio 的优势有 ＿＿＿＿＿＿＿＿＿＿＿＿＿＿＿＿＿＿＿＿＿＿。

5. Android Studio 项目中 src 目录中 androidTest 为 ＿＿＿＿＿、main 为 ＿＿＿＿＿、test 为 ＿＿＿＿＿。

三、简答题

1. 简要说明 Android Studio 项目目录结构和功能。

2. 如何在 Android Studio 中创建一个可运行的工程,简要说明。

项目二　闪屏导航

通过 U 酒保项目闪屏导航模块的实现,学习如何获取手机信息和软件版本信息,了解软件的更新原理,掌握软件下载机制,具有编写闪屏导航的能力。在任务实现过程中:

- 了解 Android 原生动作使用方法。
- 了解 TCP/IP 通信协议。
- 掌握 HttpURLConnection 使用方法。
- 掌握 PULL 解析方法。

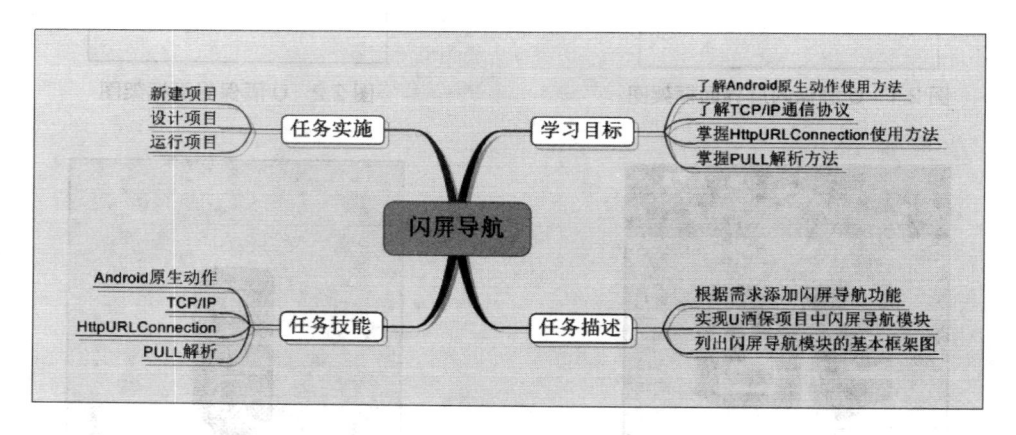

【情境导入】

开发人员在 U 酒保项目开发过程中,通过闪屏导航模块编写,实现了软件的实时更新,为用户带来更好的体验效果,引导用户熟练使用该软件。本项目实现了软件更新,Activity 动态切换等功能。讲解了 Android 原生动作的用法、TCP/IP 协议相关知识、使用 HttpURLConnection 实现软件更新以及 PULL 解析技术等。

【功能描述】

本项目将实现 U 酒保项目中闪屏导航模块：

- 实现闪屏动画。
- 实现软件版本的监测与更新。
- 实现 Activity 的动态切换。

【基本框架】

基本框架如图 2.1 和图 2.2 所示。通过本模块的学习，能将框架图 2.1 转换成效果图 2.3 所示，将框架图 2.2 转换成效果图 2.4 至图 2.6 所示。

图 2.1　U 酒保闪屏界面框架图

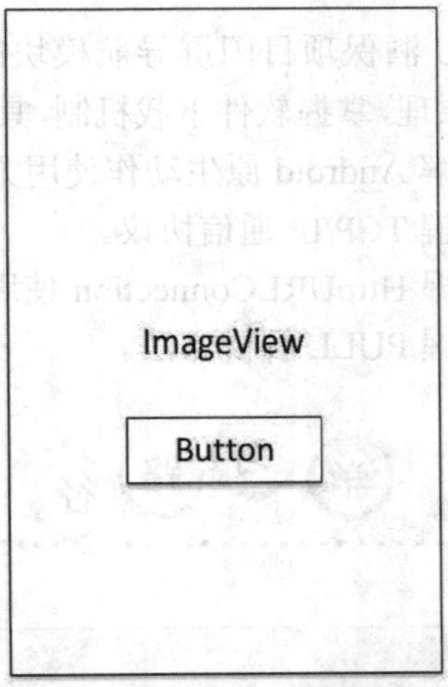

图 2.2　U 酒保界面框架图

图 2.3　U 酒保闪屏界面效果图

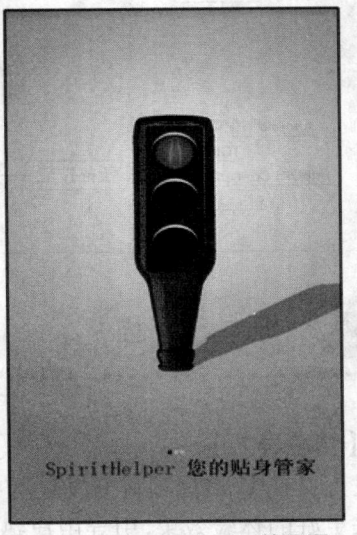

图 2.4　U 酒保导航界面效果图 1

图 2.5　U 酒保导航界面效果图 2

图 2.6　U 酒保导航界面效果图 3

技能点 1　Android 原生动作

1　原生动作简介

Android 原生动作由 Google 公司发布,不经过第三方修改。在项目开发过程中,需创建隐式 Intent 启动应用程序内的 Activity 或 SubActivity,并使用其类中的静态字符串常量(原生动作)。如表 2.1 所示。

表 2.1　Android 静态字符串常量

静态字符串常量	含　义
ACTION_ALL_APPS	用处理器打开一个列出所有已安装应用程序的 Activity
ACTION_ANSWER	用本地电话拨号程序打开一个处理来电的 Activity
ACTION_BUG_REPORT	用本地 bug 报告机制处理显示一个可以报告 bug 的 Activity
ACTION_CALL	用 Intent 的数据 URI 提供的号码拨打电话,打开一个电话拨号程序
ACTION_CALL_BUTTON	调用拨号 Activity 在用户按下硬件的"拨打按钮"时触发
ACTION_DELETE	打开删除 Intent 的数据 URL 中指定的数据的 Activity

续表

静态字符串常量	含　义
ACTION_DIAL	由本地 Android 电话拨号程序进行处理的,要拨打的号码由 Intent 的数据 URI 预先提供,没有直接打出
ACTION_EDIT	请求一个可以编辑 Intent 的数据 URI 预先提供数据的 Activity
ACTION_INSERT	在 Intent 的 URI 指定的游标中插入数据
ACTION_PICK	启动一个子 Activity,选择 Intent 中 URI 指定的 Content Provider 的某项
ACTION_SEARCH	用于启动特定的搜索 Activity,可以使用 SearchManager.QUERY 键把搜索词作为可以在 Intent 的 extra 中提供
ACTION_SEARCH_LONG_PRESS	由系统处理的硬件搜索键长按操作
ACTION_SENDTO	启动一个 Activity 来向 Intent 中的 URI 所指定的联系人发送短信
ACTION_SEND	启动一个 Activity 来向 Intent 中发送指定数据
ACTION_VIEW	通用动作,以合理的方式查看 Intent 中 URI 的内容
ACTION_WEB_SEARCH	打开浏览器,根据 SearchManager.QUERY 键指定搜索内容

2　Intent 动作机制

（1）Intent 动作机制简介

Intent 用于应用程序之间的通讯和应用程序内部的 Activity/Service 之间的交互。因此,可将 Intent 理解为不同组件之间通信的"媒介"。它可以启动一个 Activity 或启动一个服务（Service）,还可发起一个广播（Broadcasts）。Intent 通过以上方式负责对应用中单次操作的动作、动作涉及的数据、附加数据进行描述。Android 根据该 Intent 的描述,负责找到对应的组件,将 Intent 传递给被调用的组件,完成组件调用。

（2）Intent 启动组件的方法

Intent 启动组件方法分为显式启动和隐式启动。

● 显示启动

显式启动 Activity 类,创建一个新的 Intent 对象,指定当前 Activity 的上下文以及准备启动 Activity 的类。将 Intent 发送至 MyOtherActivity,具体代码如下所示:

```
Intent intent = new Intent(MyActivity.this,MyOtherActivity.class);
    startActivity(intent);
```

● 隐式启动

隐式启动 Activity 类,系统启动一个可执行一定动作的 Activity,具体代码如下所示:

```
Intent intent = new Intent(Intent.ACTION_DIAL,Uri.parse("tel:555-2368"));
    startActivity(intent);
```

3 原生动作实现步骤

实现照片选择器 demo，使用原生动作调用系统摄像头拍照，并将照片显示到界面，还可进行照片选择，效果如图 2.7 所示。实现步骤如下所示。

（1）复制图片工具类到如图 2.8 所示路径中。

图 2.7　照片选择器主界面

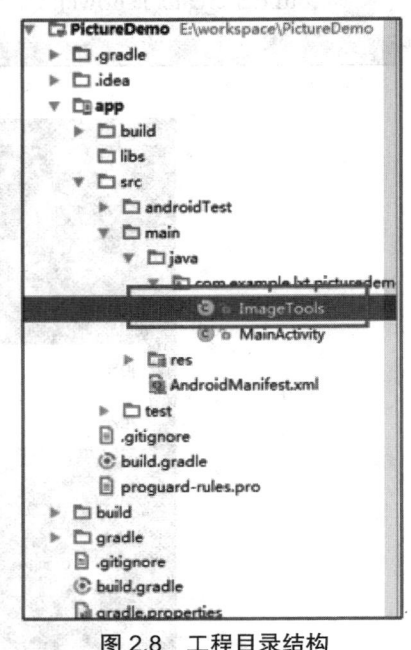

图 2.8　工程目录结构

（2）初始化界面，实现"选择拍照或相册"按钮单击事件，并弹出拍照相册选择框，效果如图 2.9 所示，具体代码如 CORE0201 所示。

```
代码 CORE0201  按钮单击选择
iv_image = (ImageView) this.findViewById(R.id.img);
            this.findViewById(R.id.btn).setOnClickListener(new OnClickListener() {
                @Override
                publicvoid onClick(View v) {
                    showPicturePicker(MainActivity.this, false);      // 即拍即显示
                } });
publicvoid showPicturePicker(Context context, boolean isCrop) {
            finalboolean crop = isCrop;
            AlertDialog.Builder builder = new AlertDialog.Builder(context);
            builder.setTitle(" 图片来源 ");
            builder.setNegativeButton(" 取消 ", null);
            builder.setItems(new String[] { " 拍照 ", " 相册 "},
                            new DialogInterface.OnClickListener() {
```

```
                                         int REQUEST_CODE;              // 类型码
                                         @Override
                                         publicvoid onClick(DialogInterface dialog, int which) {
                                         }  });
                        builder.create().show();
       }
```

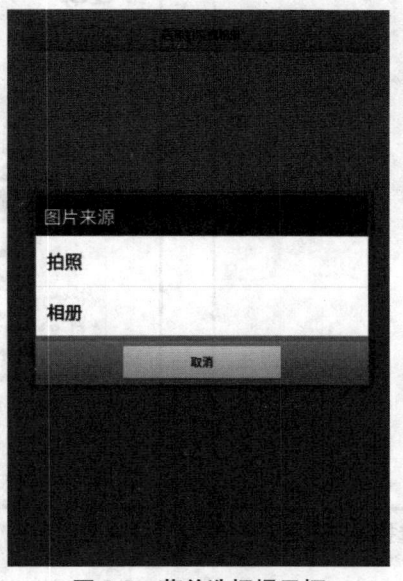

图 2.9 菜单选择提示框

（3）选择拍照或相册，调用 Android 原生动作进入系统拍照或相册界面，进行拍照并显示到主界面中或选择相册显示到主界面，效果如图 2.10 至图 2.12 所示，具体如代码如CORE0202 所示。

```
代码 CORE0202 进入拍照或相册界面

public void showPicturePicker(Context context, boolean isCrop) {
    finalboolean crop = isCrop;
    AlertDialog.Builder builder = new AlertDialog.Builder(context);
    builder.setTitle(" 图片来源 ");
    builder.setNegativeButton(" 取消 ", null);
    builder.setItems(new String[] { " 拍照 ", " 相册 " },
            new DialogInterface.OnClickListener() {
                int REQUEST_CODE;              // 类型码
                @Override
    public void onClick(DialogInterface dialog, int which) {
        switch (which) {
```

```java
caseTAKE_PICTURE:
    Uri imageUri = null;
    String fileName = null;
    Intent openCameraIntent = new Intent(
    MediaStore.ACTION_IMAGE_CAPTURE);
if (crop) {
    REQUEST_CODE = CROP;
SharedPreferences sharedPreferences = getSharedPreferences
    ("temp", Context.MODE_WORLD_WRITEABLE);
// 删除上一次截图的临时文件
ImageTools.deletePhotoAtPathAndName(
Environment.getExternalStorageDirectory().getAbsolutePath(),
sharedPreferences.getString("tempName", ""));
// 保存本次截图临时文件名字
    fileName = String.valueOf(System.currentTimeMillis()) + ".jpg";
    Editor editor = sharedPreferences.edit();
    editor.putString("tempName", fileName);
        editor.commit();
        } else {
        REQUEST_CODE = TAKE_PICTURE;
        fileName = "image.jpg";.
        }
/*
    指定照片保存路径（SD 卡），image.jpg 为一个临时文件，每次拍照后这个图片都
会替换
    */
    imageUri = Uri.fromFile(new File(Environment.getExternalStorageDirectory(), file-
Name));
            openCameraIntent.putExtra(MediaStore.EXTRA_OUTPUT,imageUri);
        startActivityForResult(openCameraIntent,REQUEST_CODE);
                break;
                caseCHOOSE_PICTURE:
                    Intent openAlbumIntent = new Intent(
                    Intent.ACTION_GET_CONTENT);
                if (crop) {
                        REQUEST_CODE = CROP;
                } else {
                        REQUEST_CODE = CHOOSE_PICTURE;
```

```
                                    }
openAlbumIntent.setDataAndType(
MediaStore.Images.Media.EXTERNAL_CONTENT_URI,"image/*");
            startActivityForResult(openAlbumIntent,REQUEST_CODE);
                    break;
            } } });
            builder.create().show();
    }
```

图 2.10　系统拍照图

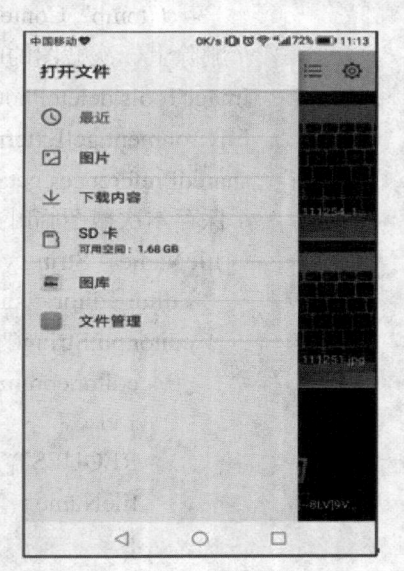

图 2.11　系统相册

图 2.12　照片选择器主界面

拓展 至此，对 Android 原生动作的介绍就要画上句号了，就知道爱学习的你们一定意犹未尽，这里就不卖关子了。想要了解原生动作的实现方法以及更多的原生动作常量，请扫描下方的二维码吧。有你想不到的惊喜哦！

技能点 2 TCP/IP

1 TCP/IP 简介

TCP 是一种可靠的连接传送服务。在传送过程中，应用程序计算机之间通过比特流通信（即数据作为无结构字节流）建立会话，其过程分以下三个阶段：

（1）通过 TCP 传输的字节流指定顺序号，获得可靠性。

（2）TCP 使用 IP 提供的网络互联功能进行稳定的数据传输。

（3）IP 不断将报文传到网络上，TCP 负责确认报文是否上传。在协同 IP 操作中 TCP 负责握手过程、报文管理、流量控制、错误检测及处理。

2 TCP/IP 可靠性

TCP/IP 的可靠性可以从以下六个方面体现：

（1）应用程序被分割为 TCP 最为合适发送的数据块，由 TCP 传递给 IP 的信息单位，称为报文段或段（segment）。

（2）TCP 发出一个报文段后，会启动一个定时器，等待目的端确认收到报文段。如不能及时收到确认，将重发该报文段。当 TCP 收到发自 TCP 连接另一端数据时，将发送一个确认。TCP 有延迟确认的功能，此功能未打开时是立即确认，打开后，由定时器触发确认时间点。

（3）TCP 将保持首部和数据的检验和。是一个端到端的检验和，目的是检测数据在传输过程中发生变化。如果收到报文段的检验和差错，TCP 将丢弃这个报文段并且不确认收到此报文段。

（4）TCP 报文段作为 IP 数据报传输，而 IP 数据报的到达可能会失序，因此 TCP 报文段的到达也可能会失序。TCP 将对收到的数据进行重新排序，以正确的顺序交给应用层。

（5）IP 数据报会发生重复，TCP 的接收端会丢弃重复数据。

（6）TCP 可提供流量控制，TCP 连接的每个对象都有固定大小的缓冲空间。TCP 的接收端只允许另一端发送接收端缓冲区所能接纳的数据。这将防止较快主机导致较慢主机的缓冲区溢出。

3 TCP 首部结构

TCP 首部长度单位为 4 字节，可表示的最大十进制值是 15，该字段的单位是 32 位字节。当 IP 首部长度为 1111（就是十进制 15），首部长度达到最大值 60 字节。TCP 首部结构如表 2.2 所示。

Android 模块化项目实战

表 2.2　TCP 首部结构

字　段	含　义
16 位源端口号	指建立连接（或发送数据）的端口号
16 位目的端口号	指连接另一端（或接受数据）的端口号
32 位序号	发送的字节序号，如果是新建立的连接，则第一个包的 seq 为 0，否则为上一个数据包的确认序号。同一个包中的序号和确认序号是不同的
32 位确认序号	等于接收到数据包的序号 seq+ 数据包的长度 len。同时告诉对端，下一个数据包的开头字节序号
4 位数据偏移	TCP 包首部的长度
保留（6 位）	为了将来定义新的用途而保留的位，但是目前置为 0

拓展　通过本节的学习，我们已经对 TCP 的功能有了一定的了解。但是 TCP 的历史和它的背景想必你们还不是很清楚。想知道更多的 TCP 的奥秘吗？扫描下方二维码为你揭开 TCP 神秘的面纱。

技能点 3　HttpURLConnection

1　HTTP 头字段简介

HTTP 头字段指在 HTTP 请求和回复消息中协议头部分的组件。它定义了某个 HTTP 事务中的操作参数。当建立 HttpURLConnection 和远程资源连接时，程序可以通过如表 2.3 所示方法设置请求头字段。

表 2.3　请求头字段方法

请求方法	含　义
setAllowUserInteraction (boolean allowuserinteraction)	设置该 URLConnection 的 AllowUserInteraction 请求的头字段值
SetDoInput(boolean doinput)	设置该 URLConnection 的 DoInput 请求字段的值
setDoOutput(boolean doinput)	设置该 URLConnection 的 DoOutput 请求字段的值
setIfModifiedSince (boolean ifModifiedSince)	设置该 URLConnection 的 IfModifiedSina 请求头字段的值
setUseCaches (boolean ifModifiedSince)	设置该 URLConnection 的 UseCaches 请求头字段的值

通过设置请求头字段建立远程资源连接后，程序可使用如表 2.4 所示方法访问头字段和内容。

项目二　闪屏导航

表 2.4　访问头字段和内容方法

方　　法	含　　义
Object getContent(String)	获取 URLConnection 内容
String getHeaderField(String name)	获取指定相应头字段值
getInputStream(String)	返回 URLConnection 对应的输入流,获取响应内容
getOnputStream(String)	返回 URLConnection 对应的输出流,发送请求参数

访问头字段后,可以使用如表 2.5 所示方法获取特定响应头字段值。

表 2.5　获取特定响应头字段值方法

方法	含义
getContentEncoding(String)	获取 contnet-encoding 响应头字段值
getContentLength(int)	获取 content-length 响应头字段值
getContentType(String)	获取 content-type 响应头字段值
getDate(datetime)	获取 date 响应头字段值
getExpiration(String)	获取 expires 响应头字段值
getLastModified(String)	获取 last-modified 响应头字段值

2　URL 简介

URL 对象是统一资源定位符,由协议名、主机、端口和资源组成,是互联网上标准资源的地址。资源可以是简单的文件或目录,也可以是复杂对象的引用。互联网上每个文件都有唯一的 URL。

URL 请求分为两类:GET 请求和 POST 请求。

● GET 请求可获取静态页面,将参数放在 URL 字串后。

● POST 的参数不放在 URL 字串中,而是放在 HTTP 请求的正文中。

3　HttpURLConnection 简介

HttpURLConnection 继承了 URLConnection,URL 可传给构造器 String 类型的参数生成指向特定地址的 URL 实例。HttpURLConnection 主要用于通过 HTTP 协议向服务器发送请求,并可以获取服务器返回的数据。

HttpURLConnection 类没有公开的构造方法,但可通过 java.net.URL 的 openConnection() 方法获取一个 URLConnection 的实例,每个 HttpURLConnection 都可用于生成单个请求,请求后在 HttpURLConnection 的 InputStream 或 OutputStream 上调用 close() 方法释放网络资源。HttpURLConnection 请求流程如图 2.13 所示。

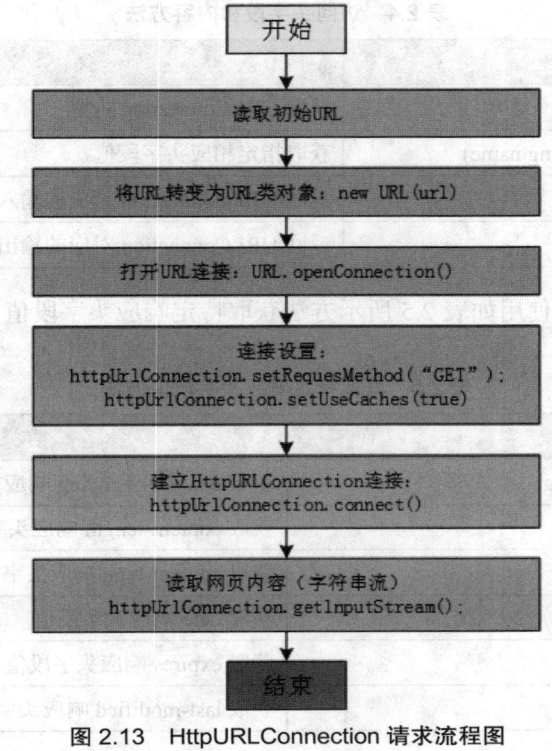

图 2.13　HttpURLConnection 请求流程图

4　HttpURLConnection 实现步骤

HttpURLConnection 运行效果如图 2.14 所示。

图 2.14　HttpURLConnection 运行效果

项目二 闪屏导航   37

（1）创建并获取 URL 地址，建立 HttpURLConnection 对象，具体代码如 CORE0203 所示。

> **代码 CORE0203 创建 HttpURLConnection 对象**
>
> URL url = new URL("http://localhost:8080/xxx.do");
> URLConnection rulConnection = url.openConnection();
> /*
> 此处的 urlConnection 对象实际上是根据 URL 的请求协议（此处是 http）生成的
> URLConnection 类的子类 HttpURLConnection，故此处最好将其转化为 HttpURLCon-
> nection 类型的对象，以便用到 HttpURLConnection 更多的 API
> */
> HttpURLConnection httpUrlConnection = (HttpURLConnection) rulConnection;

（2）设置 HttpURLConnection 为 POST 请求方式，判断 httpUrlConnection 的读入方式，并
连接服务器。具体代码如 CORE0204 所示。

> **代码 CORE0204 设置 HttpURLConnection 参数**
>
> // 设定请求的方法为 POST，默认是 GET
> httpUrlConnection.setRequestMethod("POST");
> /*
> 设置是否向 httpUrlConnection 输出，因为这个是 POST 请求，参数要放在 HTTP 正
> 文内，因此需要设为 true，默认情况下是 false
> */
> httpUrlConnection.setDoOutput(true);
> // 设置是否从 httpUrlConnection 读入，默认情况下是 true
> httpUrlConnection.setDoInput(true);
> //POST 请求不能使用缓存
> httpUrlConnection.setUseCaches(false);
> /*
> 设定传送的内容类型是可序列化的 Java 对象（如果不设此项，在传送序列化对象
> 时，当 Web 服务默认的不是这种类型时可能抛 java.io.EOFException）
> */
> httpUrlConnection.setRequestProperty("Content-type","application/x-java-serial-
> ized-object");
> // 连接，从上述 url.openConnection() 至此的配置必须要在 connect() 之前完成
> httpUrlConnection.connect();

（3）建立 URLConnection 连接，具体代码如 CORE0205 所示。

代码 CORE0205　建立 URLConnection 连接
*/ 此处 getOutputStream() 会隐含的进行 connect（即：如同调用上面的 connect() 方法在开发中不调用上述的 connect() 也可以） */ OutputStream outStrm = httpUrlConnection.getOutputStream();

（4）创建输出流对象，写入数据，发送 HttpURLConnection 请求，具体代码如 CORE0206 所示。

代码 CORE0206　发送 HttpURLConnection 请求

// 现在通过输出流对象构建对象输出流，以实现输出可序列化的对象
ObjectOutputStream objOutputStrm = new ObjectOutputStream(outputStream);
// 向对象输出流写出数据，这些数据将存到内存缓冲区中
objOutputStrm.writeObject(new String(" 我是测试数据 "));
// 刷新对象输出流，将任何字节都写入潜在的流中（此处为 ObjectOutputStream）
objOutputStm.flush();
*/
关闭流对象。不能再向对象输出流写入任何数据，先前写入的数据存在于内存缓冲区中在调用下边的 getInputStream() 函数时把准备好的 HTTP 请求正式发送到服务器
*/
objOutputStm.close();

（5）HttpURLConnection 获取响应，具体代码如 CORE0207 所示。

代码 CORE0207　HttpURLConnection 获取响应
// 调用 HttpURLConnection 连接对象的 getInputStream() 函数 InputStream inStrm = httpConn.getInputStream();

（6）将接受到的信息转换为字符串，并输出结果。具体代码如 CORE0208 所示。

代码 CORE0208　将接受到的信息转换位字符串
// 将接受到的流转换为字符串形式并输出 OutputStream os = httpConn.getOutputStream(); String param = new String(); System.out.println(param);

拓展　同学你知道吗？Android 对 HTTP 提供了很好的支持，包括两种接口，一是上边介绍的 HttpURLConnection，还有就是 Apache 接口——HttpClient。诱惑够不够大？扫描下方二维码来满足你！

技能点 4　PULL 解析

1　XML 简介

XML（可拓展标记语言）提供描述结构化数据的方法是一种简单、与平台无关并被广泛采用的标准。XML 相对于 HTML 的优点是它将用户界面与结构化数据分隔开来。使得集成来自不同源的数据成为可能。客户信息、订单、研究结果、账单付款、病历、目录数据及其他信息都可以转换为 XML。与控制数据的显示和外观的 HTML 标记不同，XML 标记用于定义数据本身的结构和数据类型。

2　PULL 简介

Android 移动设备资源宝贵，内存有限，根据不同需求选择以下技术来解析 XML，有利于提高访问的速度。

● PULL 解析器的运行方式基于事件模式，PULL 在解析过程中需自己获取产生的时间并做出相应的操作。PULL 解析器具有小巧轻便、解析速度快、简单易用等特点，适用于 Android 移动设备，Android 系统内部使用 PULL 解析器解析各种 XML。

● DOM 解析器具有简单、直观等特点。在处理 XML 文件时，将 XML 文件解析成树状结构并放入内存中进行处理。在 XML 文件较小时，可选择该解析器。

● SAX 解析器是以事件作为解析 XML 文件的模式，它将 XML 文件转化成一系列的事件，由不同的事件处理器来决定如何处理。XML 文件较大时，选择 SAX 技术较为合理。SAX 技术可高效利用 Android 有限的内存资源，并且 Android 提供了传统的 SAX 使用方法以及便捷的 SAX 包装器。

3　PULL 解析器工作原理及方法

XML PULL 在解析过程中返回数字，需要获取产生的事件，并进行相关操作如表 2.6 所示。

表 2.6　XML PULL 解析过程

事　件	返回值
读取到 XML 声明	START_DOCUMENT
结束	END_DOCUMENT
开始标签	START_TAG
结束标签	END_TAG
文本	TEXT

40　　　　　　　　　　　　Android 模块化项目实战

PULL 提供了开始元素和结束元素。当某个元素开始时，可以调用 parser.nextText 从 XML 文档中提取所有字符数据。当解析到文档结束时，自动生成 EndDocument 事件，常用的 XML PULL 接口如表 2.7 所示。

表 2.7　XML PULL 接口和类

接　口	含　义
XmlPullParser	该解析器是一个在 org.xmlpull.v1 中定义的解析功能的接口
XmlSerializer	该接口定义了 XML 信息集的序列
XmlPullParserFactory	该类用于在 XML PULL V1 API 中创建 XML PULL 解析器
XmlPullParserException	抛出单一的 XML PULL 解析器相关的错误

使用表 2.7 所示接口实现 PULL 解析 XML 文件，效果如图 2.15 所示。

操作步骤如下：

（1）创建 parseXMLWithPull() 解析 XML 文件，设置输入内容，并获取解析事件，具体代码如 CORE0209 所示。

代码 CORE0209　使用 PULL 解析 XML 文件

```
// 用 PULL 方式解析 XML
private void parseXMLWithPull(String xmlData){
try {
XmlPullParserFactory factory = XmlPullParserFactory.newInstance(); // 创建解析工厂
XmlPullParser xmlPullParser = factory.newPullParser();        // 从工厂获取解析器
// 此处填写设置输入的内容
// 此处填写获取当前解析事件,返回的是数字
// 保存内容
} catch (Exception e) {
 e.printStackTrace();
} }
```

（2）设置输入内容，具体代码如 CORE0210 所示。

代码 CORE0210　设置输入内容

```
xmlPullParser.setInput(new StringReader(xmlData));
```

（3）获取当前解析时间，返回数字，具体代码如 CORE0211 所示。

代码 CORE0211　获取当前解析时间,返回数字

```
int eventType = xmlPullParser.getEventType();
```

项目二 闪屏导航 41

（4）保存内容，具体代码如 CORE0212 所示。

代码 CORE0212 保存内容

```
String id = "007";
String name = " 小明 ";
String version="v1.1";
```

（5）判断是否结束返回，并获取 START_TAG 开始解析 XML，具体代码如 CORE0213 所示。

代码 CORE0213 解析 XML

```
// 判断是否为结束返回
while (eventType != (XmlPullParser.END_DOCUMENT)){
 String nodeName = xmlPullParser.getName();        // 获取姓名
switch (eventType){
case XmlPullParser.START_TAG:                        // 开始解析 XML
{
// 此处填写 nextText() 用于获取节点内具体内容代码
} break;
// 此处填写结束解析代码
default: break;
} }
```

（6）用 nextText() 获取节点内具体内容，具体代码如 CORE0214 所示。

代码 CORE0214 获取节点内具体内容

```
if("id".equals(nodeName)) id = xmlPullParser.nextText();
elseif("name".equals(nodeName)) name = xmlPullParser.nextText();
elseif("version".equals(nodeName)) version = xmlPullParser.nextText();
```

（7）获取结束解析标签 END_TAG，具体代码如 CORE0215 所示。

代码 CORE0215 结束解析标签 END_TAG

```
case XmlPullParser.END_TAG:
{ if("app".equals(nodeName)){
// 将解析结果进行打印显示到控制台
 Log.d(TAG, "parseXMLWithPull: id is "+ id);
 Log.d(TAG, "parseXMLWithPull: name is "+ name);
 Log.d(TAG, "parseXMLWithPull: version is "+ version);
} }
```

42　　　Android 模块化项目实战

（8）运行项目，实现如图 2.15 所示效果。

```
18423-18423/com.example.lxt.ujb_login D/ContentValues: parseXMLWithPull: id is 007
18423-18423/com.example.lxt.ujb_login D/ContentValues: parseXMLWithPull: name is 小明
18423-18423/com.example.lxt.ujb_login D/ContentValues: parseXMLWithPull: version is v1.1
```

图 2.15　PULL 解析结果

通过以下步骤实现 U 酒保的闪屏导航模块。

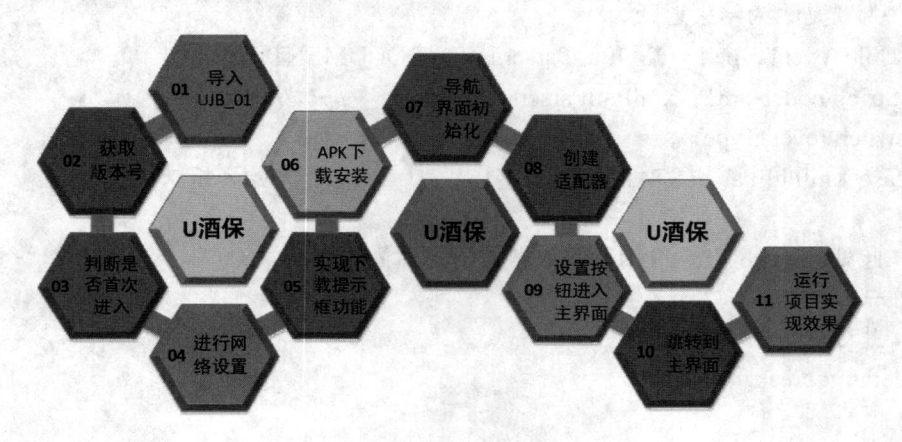

具体步骤如下所示。

第一步：将 UJB_01 导入工程，在其基础上进一步实现 UJB 项目闪屏导航模块。首先点击"Open an existing Android Studio project"打开磁盘路径查找所需项目并导入，如图 2.16 和图 2.17 所示。实现如图 2.18 所示结果图。

图 2.16　导入项目

项目二　闪屏导航

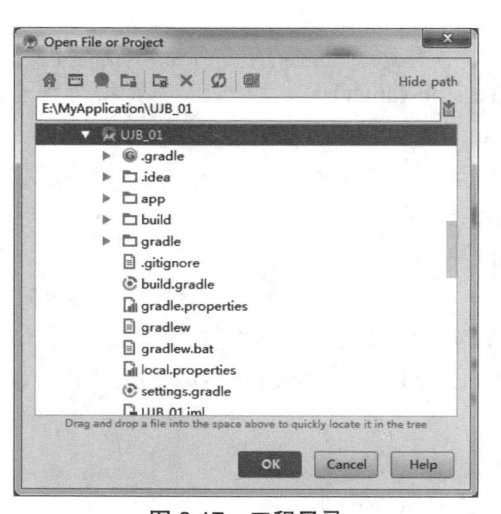

图 2.17　工程目录

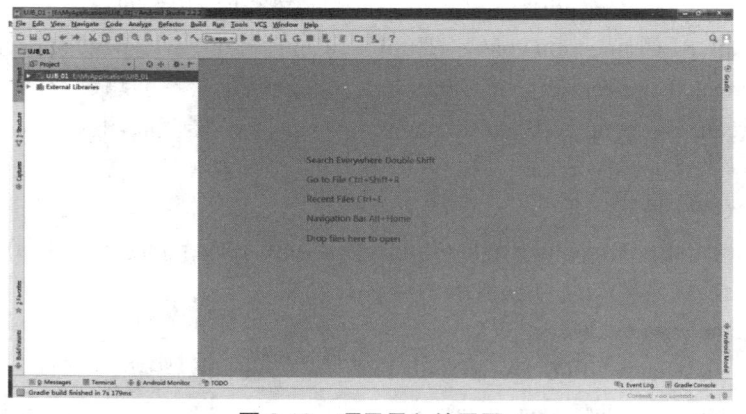

图 2.18　项目导入结果图

第二步：获取上下文、配置文件、网络管理器以及当前软件版本号，并从服务器获取最新版本号进行比对，如代码 CORE0216 所示。

代码 CORE0216　获取软件及管理者信息进行比对

```
@Override
protected void onCreate(Bundle savedInstanceState) {
    super.onCreate(savedInstanceState);
    view = View.inflate(this, R.layout.wellcom, null);
    setContentView(view);
/* 此处添加获取上下文、配置文件、网络管理器以及版本号代码 */
context = this;                                // 得到上下文
shared = new SharedConfig(context).GetConfig(); // 得到配置文件
netManager = new NetManager(context);          // 得到网络管理器
```

```
try {
localVersion = getVersionName();                     // 将得到的版本号赋给 localVersion
CheckVersionTask cv = new CheckVersionTask();
new Thread(cv).start();                              // 开启线程
} catch (Exception e) {
e.printStackTrace();
} }
/* 此处添加从服务器获取 XML 解析并进行比对版本号 */
public class CheckVersionTask implements Runnable {
  public void run() {
    try {
      // 从资源文件获取服务器址
      String path = getResources().getString(R.string.url_server);
URL url = new URL(path);                          // 包装成 URL 的对象
      HttpURLConnection conn = (HttpURLConnection) url .openConnection();
      conn.setConnectTimeout(5000);                  // 网络延时不能超过 5000 ms
InputStream is = conn.getInputStream();            // 从服务器获取输入流
String a = "" + localVersion + "";                 // 当前软件版本号
// 网络获取软件版本信息，info.getVersion() 为软件版本号
info = UpdataInfoParser.getUpdataInfo(is); if (info.getVersion().equals(a)) {
// 判断当前版本号与网络获取版本号是否一致
Message msg = new Message();
        msg.what = UPDATA_NONEED;      // 如果一致则抛出 UPDATA_NONEED
        handler.sendMessage(msg);
      } else {                         // 如果不一致则抛出 UPDATA_CLIENT
        Message msg = new Message();
        msg.what = UPDATA_CLIENT;
        handler.sendMessage(msg);
      } catch (Exception e) {          // 如果有异常则抛出 UPDATA_NONEED
        Message msg = new Message();
        msg.what = GET_UNDATAINFO_ERROR;
        handler.sendMessage(msg);
        e.printStackTrace();
    } } }
  Handler handler = new Handler() {
  @Override
  public void handleMessage(Message msg) {
    super.handleMessage(msg);
```

项目二　闪屏导航

```
    switch (msg.what) {                // 版本一致,不需要更新
      case UPDATA_NONEED:
        into() ;
        break;
      case UPDATA_CLIENT:              // 版本不一致,需要更新版本
        showUpdataDialog();
        break;
  case GET_UNDATAINFO_ERROR:          // 出现异常
        into() ;
        break; } } };
```

第三步:如果软件不需要更新,进行"第一次"进入软件判断,如果是第一次安装,则进入导航界面,否则进入主界面,并设置闪屏动画,如代码 CORE0217 所示。效果如图 2.19 所示。

代码 CORE0217　判断是否第一次安装软件并设置闪屏动画

```
private void into() {
  if (netManager.isOpenNetwork()) {       // 判断网络是否打开
FirstIntent();                            // 第一次进入软件
  } }
private void FirstIntent() {
/* 此处添加闪屏动画功能 */
// 如果网络可用则判断是否第一次进入,如果是第一次则进入欢迎界面
first = shared.getBoolean("First", true);
// 设置动画效果是 alpha,在 anim 目录下的 alpha.xml 文件中定义动画效果
animation = AnimationUtils.loadAnimation(this, R.anim.alpha);
view.startAnimation(animation);           // 给 view 设置动画效果
  animation.setAnimationListener(new AnimationListener() {
      @Override
      public void onAnimationStart(Animation arg0) {}
      @Override
      public void onAnimationRepeat(Animation arg0) {}
/* 这里监听动画结束的动作,在动画结束的时候开启一个线程,这个线程中绑定一
个 Handler,并在这个 Handler 中调用 goHome 方法,而通过 postDelayed 方法使这个方
法延迟 500ms 执行,达到持续显示第一屏 500ms 的效果 */
      @Override
      public void onAnimationEnd(Animation arg0) {
      new Handler().postDelayed(new Runnable() {
          @Override
```

```
        public void run() {
            Intent intent;
            if (first) {                    // 如果第一次,则进入引导页 WelcomeActivity
    intent = new Intent(WelcomActivity.this,GuideActivity.class);
            } else {                        // 如果不是第一次进入,则跳转到主界面
    intent = new Intent(WelcomActivity.this, MainActivity.class);
            }
            startActivity(intent);        // 开启跳转
            overridePendingTransition(R.anim.in_from_right, R.anim.out_to_left);
                                          // 设置 Activity 的切换效果
            WelcomActivity.this.finish();    }
        }, TIME);
    } }); }
```

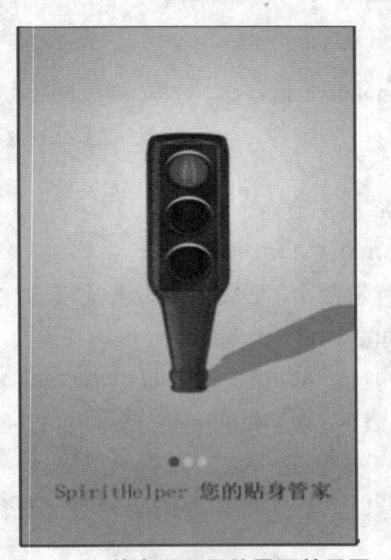

图 2.19   首次进入导航界面效果图

第四步:判断当前网络连接状态,如果没有设置网络时,进行网络设置。如代码 CORE0218 所示,效果如图 2.20 所示。

代码 CORE0218 设置网络

```
private void into() {
  if (netManager.isOpenNetwork()) {
    FirstIntent();
  } else {
/* 4 此处添加网络设置代码 */
```

```java
        builder.setTitle(" 没有可用的网络 ");
    // 如果网络不可用, 则弹出对话框, 对网络进行设置
        Builder builder = new Builder(context);
        builder.setMessage(" 是否对网络进行设置 ?");
        builder.setPositiveButton(" 确定 ",
            new android.content.DialogInterface.OnClickListener() {
              @Override
        public void onClick(DialogInterface dialog, int which) {
                Intent intent = null;
                try {
                    String sdkVersion = android.os.Build.VERSION.SDK;
                    if (Integer.valueOf(sdkVersion) > 10) {
                        intent = new Intent(                              // 获取系统设置服务
                        android.provider.Settings.ACTION_WIRELESS_SETTINGS);
                    } else {
    // 通历组件名称根据需打开响应组件
                        intent = new Intent();
                        ComponentName comp = new ComponentName(
                        "com.android.settings",
                        "com.android.settings.WirelessSettings");
                        intent.setComponent(comp);
                        intent.setAction("android.intent.action.VIEW"); // 打开网络设置界面
                        }
                    WelcomActivity.this.startActivity(intent);
                    } catch (Exception e) {
                    e.printStackTrace();
                    }  }  });
        builder.setNegativeButton(" 取消 ", new android.content.DialogInterface.OnClick-
Listener() {
                        @Override
            public void onClick(DialogInterface dialog, int which) {
            // 如果网络可用则判断是否第一次进入, 如果是第一次则进入欢迎界面
            FirstIntent(); }
        });builder.show();
    } }
```

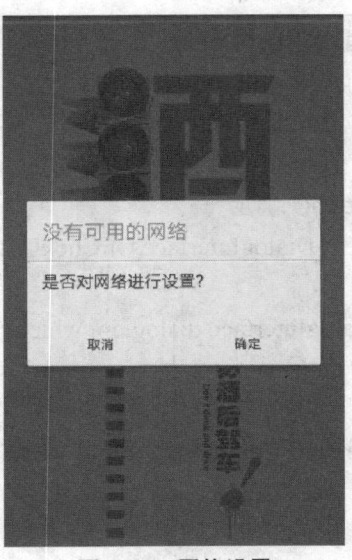

图 2.20　网络设置

　　第五步：如果当前版本号小于服务器获取版本号，实现下载提示框功能，如代码 CORE0219 所示，效果如图 2.21 所示。

```
代码 CORE0219　下载最新软件安装包
/* 此处添加下载提示框代码 */
protected void showUpdataDialog() {
  AlertDialog.Builder builer = new Builder(this);
  builer.setTitle(R.string.info);                        // 标题
  builer.setIcon(R.drawable.updateicon);                 //Logo
  builer.setMessage(info.getDescription());
// 当点确定按钮时从服务器上下载新的 APK 然后安装
  builer.setPositiveButton(" 确定 ", new DialogInterface.OnClickListener() {
public void onClick(DialogInterface dialog, int which) {
downLoadApk();                                           // 实现 APK 下载
  } });
// 当点取消按钮时进行登录
  builer.setNegativeButton(" 取消 ", new DialogInterface.OnClickListener() {
  public void onClick(DialogInterface dialog, int which) {
    Message msg = new Message();
    msg.what = UPDATA_NONEED;
    handler.sendMessage(msg);
  } });
    AlertDialog dialog = builer.create();
    dialog.setCanceledOnTouchOutside(false);
```

项目二　闪屏导航

```java
        dialog.show();                                      // 显示提示框
    }
    public void handleMessage(Message msg) {
        super.handleMessage(msg);
        switch (msg.what) {
            case SDCARD_NOMOUNTED:                  // SD 卡不可用
                Toast.makeText(getApplicationContext(), "SD 卡不可用 ", Toast.LENGTH_
SHORT).show();
            break;
            case DOWN_ERROR:                        // 下载新版本失败
                Toast.makeText(getApplicationContext(), " 下载新版本失败 ", Toast.LENGTH_
SHORT).show();
                break;
    } } };
```

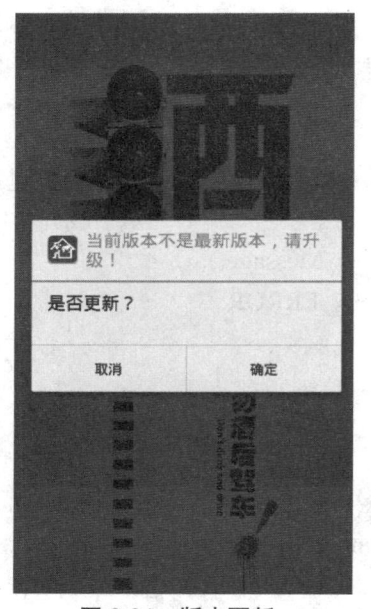

图 2.21　版本更新

第六步:判断是否有 SD 卡,并获取下载路径,通过网络实现 APK 下载功能,并安装,如代码 CORE0220 所示,效果如图 2.22 所示。

代码 CORE0220　APK 下载

```java
/* 此处添加从服务器中下载 APK 的代码 */
protected void downLoadApk() {
    final ProgressDialog pd;                         // 进度条对话框
```

```java
        pd = new ProgressDialog(WelcomActivity.this);
        pd.setProgressStyle(ProgressDialog.STYLE_HORIZONTAL);
        pd.setMessage(" 正在下载更新 ");
                                              // 判断 SD 卡是否存在
        if ( !Environment.getExternalStorageState().equals( Environment.MEDIA_MOUNT-
ED)) {
    Message msg = new Message();
        msg.what = SDCARD_NOMOUNTED; //SD 卡不存在
    handler.sendMessage(msg);
        } else {
        pd.show();                              //SD 卡存在
        new Thread() {
        @Override
        public void run() {
         try {
                                              // 获取下载路径
            File file = DownLoadManager.getFileFromServer(info.getUrl(), pd);
            sleep(1000);
            installApk(file);                 // 下载路径
            pd.dismiss();                     // 结束掉进度条对话框
            } catch (Exception e) {           // 下载异常时执行
            Message msg = new Message();
            msg.what = DOWN_ERROR;
        handler.sendMessage(msg);
            e.printStackTrace();
            } } }.start();
        } }
    protected void installApk(File file) {        // 安装 APK
      Intent intent = new Intent();
      intent.setAction(Intent.ACTION_VIEW); // 执行动作
                                              // 执行的数据类型
                intent.setDataAndType(Uri.fromFile(file), "application/vnd.android.pack-
age-archive");
      startActivity(intent);
      }
```

项目二 闪屏导航

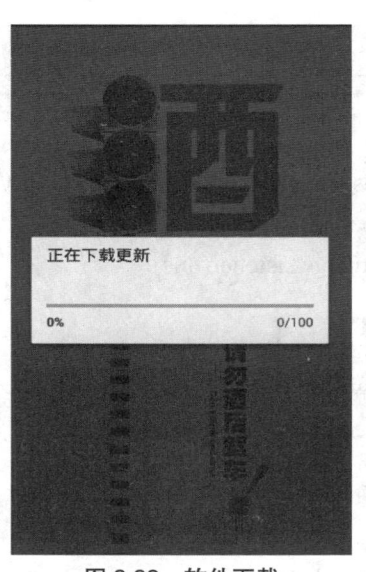

图 2.22 软件下载

第七步：通过 for 循环添加导航图片，实现导航界面初始化，如代码 CORE0221 所示。

代码 CORE0221 导航界面初始化

```
/* 此处添加导航界面舒适化的代码 */
private void inItView() {
    viewPager = (ViewPager) findViewById(R.id.viewpage);          // 初始化
    startButton = (Button) findViewById(R.id.start_Button);
    startButton.setOnClickListener(this);
    indicatorLayout = (LinearLayout) findViewById(R.id.indicator);
    views = new ArrayList<View>();
    indicators = new ImageView[images.length];                    // 添加导航图片
    for (int i = 0; i < images.length; i++) {
        ImageView imageView = new ImageView(context);
        imageView.setBackgroundResource(images[i]);               // 设置背景图片
        views.add(imageView);                                     // 将背景添加到视图中
        indicators[i] = new ImageView(context);
        indicators[i].setBackgroundResource(R.drawable.indicators_default);
        if (i == 0) {
            indicators[i].setBackgroundResource(R.drawable.indicators_now);
        }
        indicatorLayout.addView(indicators[i]);
    } }
```

第八步：创建适配器，将背景图片在 ViewPager 中进行填充，如代码 CORE0222 所示。

52　　　　　　　　　　　　Android 模块化项目实战

代码 CORE0222　创建适配器并填充信息

```java
/* 此处添加创建适配器的代码 */
private void inItView() {
  pagerAdapter = new BasePagerAdapter(views);                // 遍历 pagerAdapter 适配器
  viewPager.setAdapter(pagerAdapter);                        // 绑定适配器
  viewPager.setOnPageChangeListener(this);
}
class BasePagerAdapter extends PagerAdapter{
  private List<View> views=new ArrayList<View>();
  public BasePagerAdapter(List<View> views){
    this.views=views;
  }
  @Override
  public boolean isViewFromObject(View arg0, Object arg1) {
    return arg0 == arg1;
  }
  @Override
  public int getCount() {
    return views.size();
  }
  @Override
  public void destroyItem(View container, int position, Object object) {
    ((ViewPager) container).removeView(views.get(position));
  }
  @Override
  public Object instantiateItem(View container, int position) {
    ((ViewPager) container).addView(views.get(position));
    return views.get(position);
  } }
```

第九步：当滑动到最后一个导航界面时，显示进入 U 酒保主界面按钮，如代码 CORE0223 所示。效果如图 2.23 所示。

代码 CORE0223　显示按钮进入主界面

```java
/* 此处添加显示按钮进入主界面的代码 */
@Override
public void onPageSelected(int arg0) {
  if (arg0 == indicators.length - 1) {                // 如果是最后一个界面,显示按钮
```

项目二 闪屏导航 53

```
        startButton.setVisibility(View.VISIBLE);
    } else {                                    // 如果不是最后一个界面,隐藏按钮
        startButton.setVisibility(View.INVISIBLE);
    }for (int i = 0; i < indicators.length; i++) {
    indicators[arg0].setBackgroundResource(R.drawable.indicators_now);
    if (arg0 != i) {
    indicators[i] .setBackgroundResource(R.drawable.indicators_default);
    } }
```

图 2.23 导航最后一页效果图

第十步:点击按钮将信息传到 SharedPreferences 中进行存储,进入主界面,如代码 CORE0224 所示。

```
代码 CORE0224 跳转到主界面
/* 此处添加跳转主界面的代码 */
@Override
public void onClick(View v) {
    if (v.getId() == R.id.start_Button) {
    // 将信息传到 sp 中进行存储
    SharedPreferences shared = new SharedConfig(this).GetConfig();
    Editor editor = shared.edit();
    editor.putBoolean("First", false);
    editor.commit();
    startActivity(new Intent(GuideActivity.this, MainActivity.class));  // 跳转到主界面
```

```
        overridePendingTransition(R.anim.in_from_right, R.anim.out_to_left);
        this.finish();
    } }
```

第十一步：运行项目，实现效果如图 2.3 至图 2.6。

本项目介绍了 U 酒保闪屏导航模块的实现，通过对本项目的学习可以了解 Android 原生动作、TCP/IP 通信协议相关知识、Activity 界面切换机制，重点掌握 HttpURLConnection 使用方法及 PULL 解析方法，实现软件及时更新。

extra	额外的	answer	答案
send	发送	resource	资源
locator	定位器	protocal	会谈记要
pull	拉	down	向下
update	更新	shared	共享

一、选择题

1.Android 中文件操作模式中表示只能被本应用使用，写入文件会覆盖的是（　　）。

A.MODE_APPEND　　　　　　　　　　B.MODE_WORLD_READABLE

C.MODE_WORLD_WRITEABLE　　　　D.MODE_PRIVATE

2. 从 HTTP 请求中，获得请求参数，应该调用（　　）。

A.request 对象的 getAttribute() 方法　　　B.request 对象的 getParameter() 方法

C.session 对象的 getAttribute() 方法　　　D.session 对象的 getParameter() 方法

3. Android 解析 XML 的方法中，将整个文件加载到内存中进行解析的是（　　）。

A.SAX　　　　　B.PULL　　　　　　　　C.DOM　　　　　　　D.JSON

4. 使用 HttpUrlConnection 实现移动互联时，设置读取超时属性的方法是（　　）。

A.setTimeout()　　　　　　　　　　B.setReadTimeout()

C.setConnectTimeout()　　　　　　　D.setRequestMethod()

5. 下列不属于 PULL 解析 XML 中字段的是（　　）。

A.START_TAG　　　B.START_DOCUMENT　　　　C.NEXT_TAG　　　　D.TEXT

二、填空题

1.Android 原生动作（Native Activity）是 Intent 类中的 _____。

2.TCP 是一种较为可靠的面向连接的传送服务，传送数据的过程是分阶段的，主机与交换机间必须建立一个会话，通过 _____，即数据作为无结构字节流。

3.URL 请求分为以下两类，分别为 _____ 请求和 _____ 请求。

4.XML PULL 在解析过程中返回的是 _____。

5.DOM 解析器具有 _____、_____ 等特点。

三、上机题

1. 编写代码实现任意图片以图片中心为圆心的旋转。

2. 编写代码获取手机自带所有传感器。

项目三 登录注册

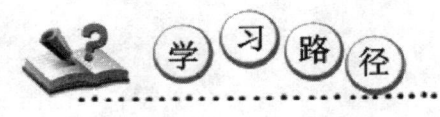

通过 U 酒保项目登录注册模块的实现,了解 SlidingMenu 侧滑功能的使用方法,掌握如何使用 ShareSDK 在应用中实现分享功能,学习 SlidingMenu 的各种属性,具有使用 SlidingMenu 侧滑功能实现侧滑效果的能力。在任务实现过程中:

● 了解 SlidingMenu 各种侧滑方式。
● 掌握 ShareSDK 使用方法。
● 掌握 SlidingMenu 的各种属性。

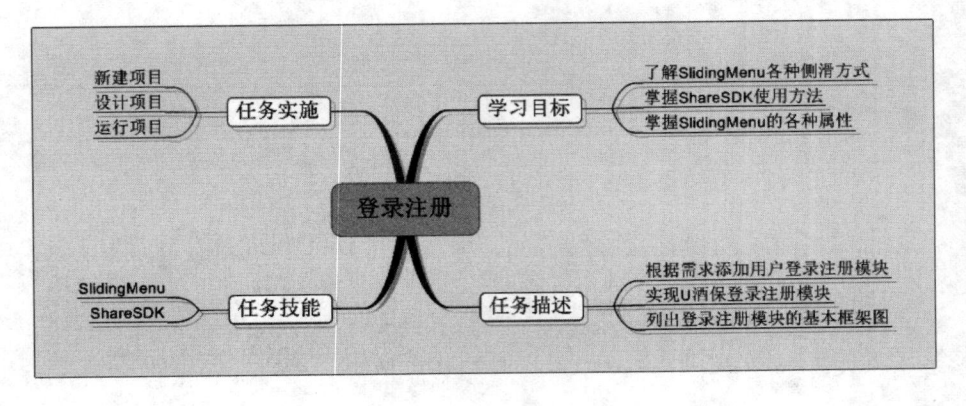

【情境导入】

在手机或电脑软件应用过程中,每款软件都会根据一些特定的方式进行用户区分,设置账号是最好的区分方式。开发人员在 U 酒保的开发过程中进行登录注册模块的开发,每个用户都可使用手机号码注册属于自己的账号,同时设置与账号匹配的密码便于后续的应用,并在此基础上添加了系统设置与好友分享等功能。

【功能描述】

本项目将实现 U 酒保登录注册模块：

- 实现 SlidingMenu 侧滑功能。
- 实现用户注册功能。
- 实现用户登录功能。
- 实现设置字体大小功能。
- 实现推送功能。
- 实现清除缓存功能。
- 实现意见反馈功能。
- 实现第三方如微信、新浪微博等好友分享功能。

【基本框架】

基本框架如图 3.1 至图 3.6 所示。通过本模块的学习,能将框架图 3.1 至图 3.6 转换成效果图 3.7 至图 3.12 所示。

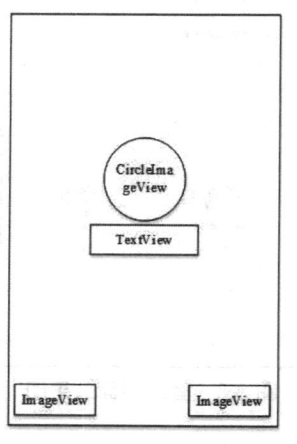

图 3.1 SlidingMenu 框架图

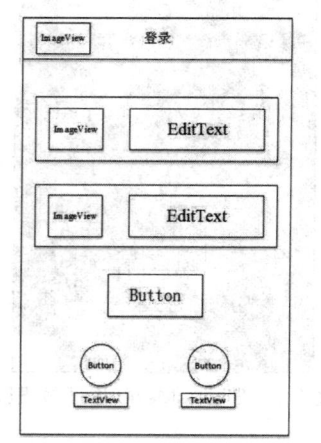

图 3.2 登录界面框架图

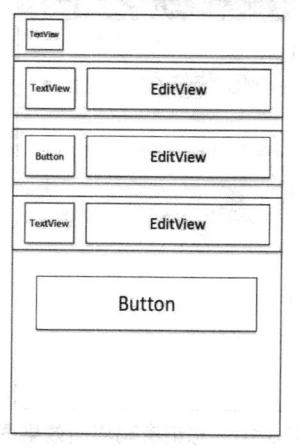

图 3.3 注册界面框架图

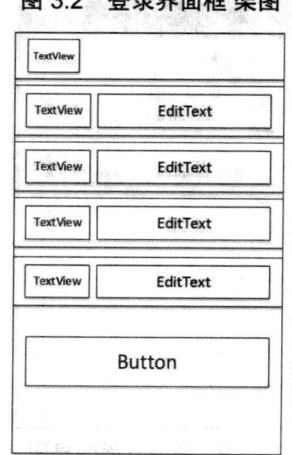

图 3.4 忘记密码界面框架图

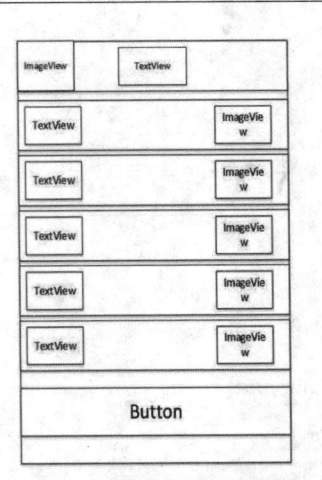

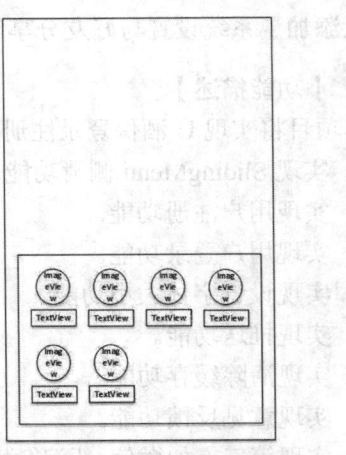

图 3.5　系统设置界面框架图

图 3.6　好友分享界面框架图

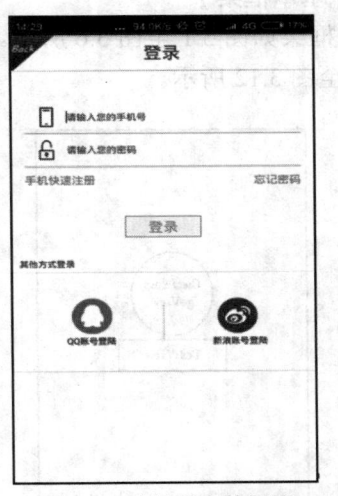

图 3.7　SlidingMenu 界面效果图

图 3.8　登录界面效果图

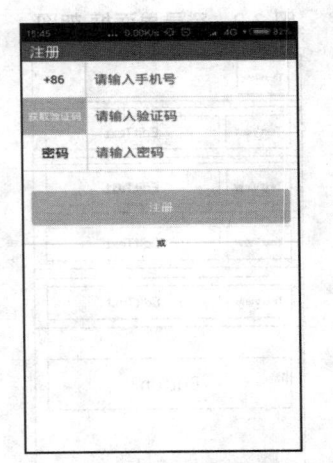

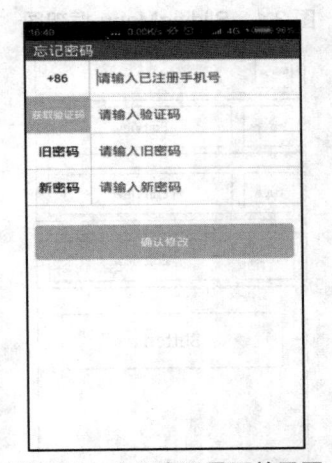

图 3.9　注册界面效果图

图 3.10　忘记密码界面效果图

项目三 登录注册 59

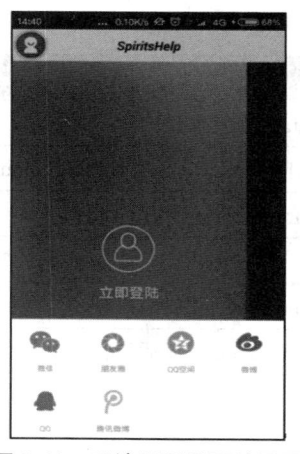

图 3.11 系统设置界面效果图

图 3.12 好友分享界面效果图

技能点 1 SlidingMenu

1 SlidingMenu 简介

SlidingMenu 定制灵活,支持各种阴影效果和渐变的滑动效果,是一种设置界面效果的新技术,在主界面左滑或者右滑时出现设置界面,方便进行系统设置、用户登录、好友分享等操作。目前有大量的应用都在使用该技术。例如：QQ 的登录模块、微信的左右页面切换、支付宝的切换功能模块等。界面显示的各种效果是通过相应的属性方法实现的,SlidingMenu 常用的属性如表 3.1 所示。

表 3.1 SlidingMenu 常用属性

属 性	含 义
menu.setMode(SlidingMenu.LEFT)	设置左滑菜单
menu.setTouchModeAbove (SlidingMenu.TOUCHMODE_FULLSCREEN)	设置滑动的屏幕范围,该设置为全屏区域都可以滑动
menu.setShadowDrawable(R.drawable.shadow)	设置阴影图片
menu.setShadowWidthRes(R.dimen.shadow_width)	设置阴影图片的宽度
menu.setBehindOffsetRes(R.dimen.slidingmenu_off-set)	SlidingMenu 划出时主页面显示的剩余宽度

续表

属　性	含　义
menu.setMenu(R.layout.menu_layout)	设置 menu 的布局文件
menu.toggle()	动态判断自动关闭或开启 SlidingMenu
menu.showMenu()	显示 SlidingMenu
menu.showContent()	显示内容
menu.OnClosedListener(OnClosedListener)	监听 SlidingMenu 关闭时事件
menu.OnClosedListener(OnClosedListener)	监听 SlidingMenu 关闭后事件

2　SlidingMenu 使用步骤

通过学习以上属性，实现侧滑菜单功能，运行效果如图 3.13 和图 3.14 所示。

图 3.13　SlidingMenu 主界面

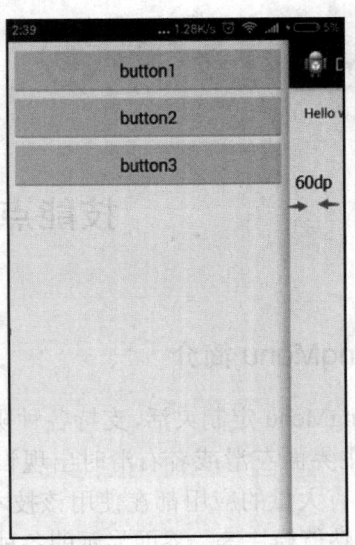

图 3.14　SlidingMenu 界面

（1）在 Activity 中通过 SlidingMenu() 构造方法，设置侧滑菜单，具体代码如 CORE0301 所示。

代码 CORE0301　创建 SlidingMenu() 构造方法

SlidingMenu menu = new SlidingMenu(MainActivity.this);

（2）根据自身需求设置侧滑菜单的参数（滑动方向、滑动方式、颜色、样式），绑定到 Activity 实现侧滑效果。具体代码如 CORE0302 所示。

代码 CORE0302　设置侧边栏的方向和滑动方式

menu.setMode(SlidingMenu.LEFT);　　　　　　　// 设置滑动方向

```java
// 设置监听开始滑动的触碰范围
public static final int TOUCHMODE_MARGIN = 0;          // 边缘
public static final int TOUCHMODE_FULLSCREEN = 1;     // 全屏
public static final int TOUCHMODE_NONE = 2;           // 不能滑动
menu.setTouchModeAbove(SlidingMenu.TOUCHMODE_MARGIN);
// 设置边缘阴影的宽度,通过 dimens 资源文件中的 ID 设置
menu.setShadowWidthRes(R.dimen.shadow_width);
// 设置边缘阴影的颜色图片,通过资源文件 ID 设置
menu.setShadowDrawable(R.drawable.shadow);
/*
设置 menu 全部打开后,主界面剩余部分与屏幕边界的距离,通过 dimens 资源文件
ID 设置
*/
menu.setBehindOffsetRes(R.dimen.slidingmenu_offset);
menu.setFadeEnabled(true); // 设置是否淡入淡出
        menu.setFadeDegree(0.35f);
/* 设置淡入淡出的值,只在 setFadeEnabled() 设置为 true 时有效 */
// 将 menu 绑定到 Activity,同时设置绑定类型
menu.attachToActivity(this, SlidingMenu.SLIDING_WINDOW);
menu.setMenu(R.layout.slide_menu);              // 设置 menu 的 layout
// 设置 menu 的背景
menu.setBackgroundColor(getResources().getColor(android.R.color.background_
dark));
    View menuroot = menu.getMenu();                    // 获取 menu 的 layout
```

技能点 2　ShareSDK

1　ShareSDK 简介

ShareSDK 是社会化分享组件,为 iOS、Android、WP8 的 APP 提供分享功能。ShareSDK 集成了一些常用的类库和接口,在缩短开发时间的同时,提供了社会化统计分析管理后台的功能,可以了解用户、信息流、回流率、传播效率等数据。它支持包括 QQ、微信、新浪微博、腾讯微博等国内外 40 多家社交平台,帮助开发者轻松实现社会化分享等功能。在 U 酒保项目中通过 ShareSDK 实现了 QQ、微博、微信快捷分享功能。ShareSDK 发展历程如表 3.2 所示。

表 3.2　ShareSDK 发展历程

时　　间	成　　果
2013 年 1 月 16 日	ShareSDK for iOS 正式发布
2013 年 3 月 18 日	ShareSDK for Android 版正式发布
2013 年 8 月	ShareSDK APP 开发者用户已超 1 万
2013 年 12 月 19 日	ShareSDK for cocos2d-x 2.2.0 专用组件正式发布
2013 年 12 月 27 日	新增评论和赞模块
2014 年 2 月 20 日	ShareSDK for Unity3d 正式发布
2014 年 3 月 10 日	ShareSDK for ANE 正式发布
2014 年 3 月 22 日	ShareSDK for Java Script 正式发布

2　ShareSDK 使用方法

（1）获取 ShareSDK

到 ShareSDK 官网注册并且创建新应用,获得 ShareSDK 的 Appkey,然后到 SDK 的下载页面下载 SDK 的压缩包。

ShareSDK 网址：http://mob.com/

（2）导入 ShareSDK

复制下载后的 SDK 包内的 jar 包和资源。在复制 jar 包的同时,不仅需要复制 MainLibs/libs 中的 jar 包,还需复制 MainLibs/res 中的图片和 strings.xml 文件,否则会出现在授权时找不到资源等问题。

（3）添加应用信息

开发者可自行选择以下三种方式添加应用信息。这三种方式优先级不同：优先级最高的方式可以实现“动态配置应用信息”的功能,但是不能脱离网络,否则 ShareSDK 无法运作。优先级中的方式是最灵活的方式,可以在代码里面写应用信息,也可以通过私有协议,从自己的服务器上动态获取应用注册信息。优先级最低的方式是最为方便、集中的。以下是三种不同优先级的使用方法。

● 优先级最高：配置在 ShareSDK 的应用管理后台中,使用此方法需要调用 ShareSDK.initSDK(context 返回的 AppKey) 方法进行初始化,更改注册信息可直接从网上更改,而不需要重新发布新版本进行更新。

● 优先级中：通过代码配置 setPlatformDevInfo(String,HashMap<String, Object>) 方法 , 使用此方法需要调用 ShareSDK.init(context,返回的 AppKey) 方法进行初始化。

● 优先级最低：通过 assets/ShareSDK.xml 文件来配置,这种方式最简单。

在 ShareSDK.xml 中配置注册信息,具体代码如下所示。

```
<ShareSDK AppKey=" 注册到的 AppKey" />
<SinaWeibo
SortId=" 平台在您分享列表中的位置,整型,数值越大越靠后 "
```

项目三　登录注册　　　63

> AppKey=" 填写您在应用上注册到的 AppKey"
> AppSecret=" 填写您在应用上注册到的 AppSecret"
> Id=" 自定义字段,整型,用于项目中对此应用的识别符 " />

　　拓展　该模块的主要任务是实现 U 酒保项目的登录注册功能。我们在写程序的同时是否曾想过登录注册的另一面?比如说对它的宏观理解、该业务中验证机制的演变以及在设计过程中的基本要求等。扫描下方二维码来解开你的疑惑。

　　通过以下步骤实现 U 酒保项目的登录注册模块。

　　具体步骤如下所示。

　　第一步:将 UJB_01 导入工程,在其基础上进一步实现 UJB 项目登录注册模块。首先点击"Open an existing Android Studio project"打开磁盘路径查找所需项目并导入,如图 3.15 和图 3.16 所示。实现如图 3.17 所示结果图。

图 3.15　导入项目

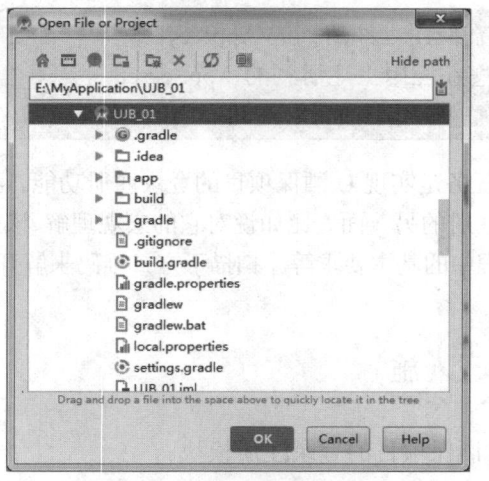

图 3.16　工程目录

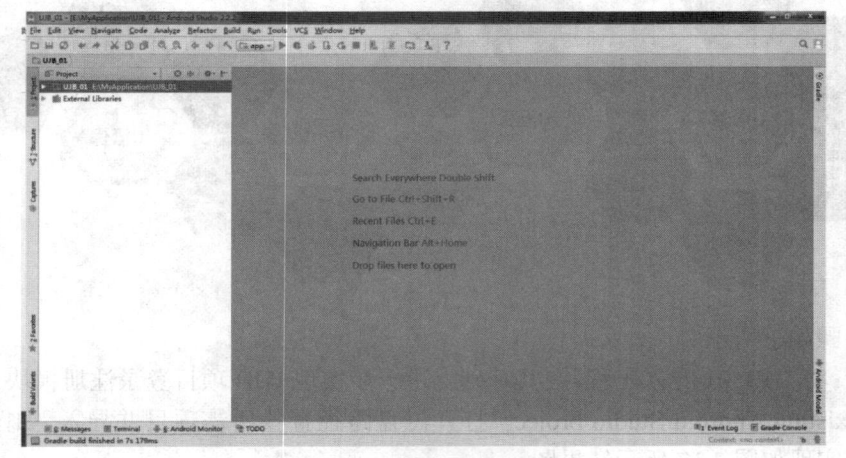

图 3.17　项目导入结果图

　　第二步：在主界面实现 SlidingMenu 属性及参数设置（滑动方向、滑动方式、颜色、样式），并初始化布局。具体如代码 CORE0303 所示。

代码 CORE0303　设置 SlidingMenu 属性参数

```
/* 此处添加设置 SlidingMenu 属性代码 */
    public class MainActivity extends Activity implements View.OnClickListener {
public static final String TAG_1 = "HOME";
public static final String TAG_2 = "ADVISE";
public static final String TAG_3 = "HELP";
public static final String TAG_4 = "INSURANCE";
private ImageView iv_left;
private CircleImageView iv_head;
```

项目三 登录注册 65

```java
private ImageView iv_setting;
private ImageView iv_share;
private TextView tv_account;
private long exitTimes;
protected SlidingMenu side_drawer;
private AppContext appContext;
private UMSocialService mController= UMServiceFactory.getUMSocialService
    ("com.umeng.share");
    @Override
protected void onCreate(Bundle savedInstanceState) {
super.onCreate(savedInstanceState);
        setContentView(R.layout.activity_tabs);
    // 消息推送
PushAgent mPushAgent = PushAgent.getInstance(this.getApplication());
    mPushAgent.enable();
    inItActionBar();                    //SlidingMenu 设置属性参数
    InItPlatform();                     // 第三方平台分享
    OncheckLogin();    }                // 监测是否登录
    // SlidingMenu 属性参数设置
private void inItActionBar() {
    ViewcustomView=LayoutInflater.from(this).inflate(R.layout.actionbar，new Lin-
earLayout(this)，false);                    // 获取第三方界面
    getActionBar().setDisplayOptions(ActionBar.DISPLAY_SHOW_CUSTOM);
    getActionBar().setCustomView(customView); // 自定义 ActionBar 布局
    iv_left = (ImageView) customView.findViewById(R.id.iv_left);
    iv_left.setOnClickListener(this);
    side_drawer = new SlidingMenu(this);
    side_drawer.setMode(SlidingMenu.LEFT);    // 设置滑动方向
    side_drawer.setTouchModeBehind(SlidingMenu.TOUCHMODE_MARGIN);
    /* 设置滑动范围 */
    side_drawer.setMenu(R.layout.slidingmenu_left); // 绑定滑出布局
    // 将抽屉菜单与主页面关联起来
    side_drawer.setBehindOffsetRes(R.dimen.slidingmenu_offset);
    side_drawer.attachToActivity(this，SlidingMenu.SLIDING_CONTENT);
    // 设置滑动时菜单的是否淡入淡出
side_drawer.setFadeEnabled(true);
side_drawer.setFadeDegree(0.4f);            // 设置淡入淡出的比例
side_drawer.setShadowWidthRes(R.dimen.shadow_width);
```

```
        iv_head = (CircleImageView) side_drawer.findViewById(R.id.iv_head);
        tv_account = (TextView) side_drawer.findViewById(R.id.tv_account);
        iv_setting = (ImageView)side_drawer.findViewById(R.id.iv_setting);
        iv_share = (ImageView)side_drawer.findViewById(R.id.iv_share);
        iv_head.setOnClickListener(this);            // 设置单击监听事件
        iv_setting.setOnClickListener(this);
        iv_share.setOnClickListener(this);
    }  }
```

第三步：点击左上方图片，实现 SlidingMenu 侧滑，具体如代码 CORE0304 所示。效果如图 3.18 所示。

图 3.18　侧滑完成后界面

代码 CORE0304　SlidingMenu 侧滑

```
/* 此处添加 SlidingMenu 侧滑代码 */
@Override
public void onClick(View v) {
  switch (v.getId()) {
  case R.id.iv_left:
  if (side_drawer.isMenuShowing()) {            // 判断 slidingMenu 是否滑出
    side_drawer.showContent();
  } else {
    side_drawer.showMenu();
} break;
```

项目三 登录注册　　　67

```
default:
break;
    } }
```

第四步：监测是否已登录，同时判断登录信息，如果未登录则跳转到登录界面，具体如代码 CORE0305 所示。

代码 CORE0305　监测是否登录

```
/* 此处添加检查是否登录代码 */
public void OncheckLogin() {
    if (appContext.isLogin() == true) {              // 判断是否登录
    if (!appContext.getHead_url().equals("")) {       // 判断头部 URL 是否为空
        iv_head.setImageBitmap(Util.getbitmap(appContext.getHead_url()));   }
        if (!appContext.getNickname().equals("")) {        // 判断昵称是否为空
            tv_account.setText(appContext.getNickname());   }
    } else {
        iv_head.setImageResource(R.drawable.login_head);
        tv_account.setText(" 立即登录 ");
    } }
```

第五步：点击"立即登录"进入登录界面，具体如代码 CORE0306 所示。

代码 CORE0306　点击"立即登录"跳转

```
/* 此处添加立即登录代码 */
@Override
public void onClick(View v) {
    switch (v.getId()) {
    case R.id.iv_head:                        // 获取点击内容 ID
    if (appContext.isLogin() == true) {
        } else {
        startActivityForResult(new Intent(MainActivity.this,LoginActivity.class), 200); }
break;
default:
break;
    } }
```

第六步：初始化注册界面控件，设置组件监听器。具体如代码 CORE0307 所示。

代码 CORE0307　初始化注册界面

```
/* 此处添加初始化注册界面代码 */
```

```
privatevoid findByView() {
        btn_back = (Button) findViewById(R.id.btn_back);        // 获取各组件 ID
        et_username = (EditText) findViewById(R.id.et_username);
        et_password = (EditText) findViewById(R.id.et_password);
        tv_register = (TextView) findViewById(R.id.tv_register);
        tv_forget = (TextView) findViewById(R.id.tv_forget);
        btn_login = (Button) findViewById(R.id.btn_login);
        btn_sina = (Button) findViewById(R.id.btn_sina);
        btn_qq = (Button) findViewById(R.id.btn_qq);
btn_back.setOnClickListener(this);                // 定义各组件监听方式
        tv_register.setOnClickListener(this);
        tv_forget.setOnClickListener(this);
        btn_login.setOnClickListener(this);
        btn_qq.setOnClickListener(this);
        btn_sina.setOnClickListener(this);
    }
```

第七步：点击"注册"按键，获取注册信息，并将注册信息添加到数据库，实现注册功能，具体如代码 CORE0308 所示。效果如图 3.19 所示。

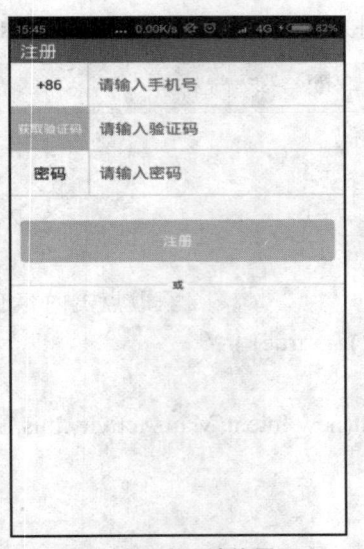

图 3.19   注册功能界面

代码 CORE0308 注册功能实现

```
/* 此处添加注册功能实现的代码 */
@Override
    public void onClick(View v) {
```

```java
                switch (v.getId()) {
                    case R.id.tv_register:
// 点击"注册"按键跳转到注册界面进行注册
startActivity(new Intent(LoginActivity.this, RegisteActivity.class));
    break;
    default:
    break;
            } }
// 注册界面
public class RegisteActivity extends Activity {
EditText et_user;
    EditText et_pwd;
    Button btn_zhuce;
    registerHelper helper = new registerHelper(this);
    @Override
    protected void onCreate(Bundle savedInstanceState) {
            super.onCreate(savedInstanceState);
                    setContentView(R.layout.activity_registe);
            et_pwd = (EditText) findViewById(R.id.et_pwd);
            et_user = (EditText) findViewById(R.id.et_user);
            btn_zhuce = (Button) findViewById(R.id.btn_zhuce);
// 给注册按钮添加监听
            btn_zhuce.setOnClickListener(new OnClickListener() {
                    @Override
            public void onClick(View arg0) {
            // TODO Auto-generated method stub
            String str_user = et_user.getText().toString().trim();
            String str_pwd = et_pwd.getText().toString().trim();
    if(str_user.length()!=11){
    Toast.makeText(RegisteActivity.this, " 输入的手机号码有误，请查证后重新输入
", 0).show();
            }else{
        if(!TextUtils.isEmpty(str_pwd)){            // 将注册信息添加到数据库中
            helper.insert(str_user, str_pwd);
            helper.close();
            Toast.makeText(RegisteActivity.this, " 注册成功 ", 0).show();
            startActivity(new Intent(RegisteActivity.this, LoginActivity.class));
            inish();
```

```
                }else{
            Toast.makeText(RegisteActivity.this, " 请将信息补充完整后进行注册 ",
0).show();
                        } } } }); } }
```

第八步：返回登录界面，输入账号密码，判断登录信息是否正确，实现用户登录，具体如代码 CORE0309 所示。

代码 CORE0309 用户登录

```
/* 此处添加用户登录代码 */
@Override
    public void onClick(View v) {
            switch (v.getId()) {
            case R.id.btn_login:                 // 判断是否为登录按钮
            String username = et_username.getText().toString().trim();
            String psd = et_password.getText().toString().trim();
        if(username.length()!=11){
        Toast.makeText(LoginActivity.this, " 输入的手机号码有误,请查证后重新输入 ",
0).show();
                    }else{
                    if(!TextUtils.isEmpty(psd)){
                            if(getMsg(username, psd)){
                            Toast.makeText(LoginActivity.this, " 登录成功 ", 0).show();
                            startActivity(new Intent(LoginActivity.this, MainActivity.class));
                            }            }}
                    break;
            default:
                    break;
            } }
```

第九步：点击"忘记密码"按键，进入到忘记密码界面，核对信息正确后，实现修改密码功能，具体如代码 CORE0310 所示。效果如图 3.20 所示。

代码 CORE0310 忘记密码功能

```
/* 此处添加忘记密码功能代码 */
public class ForgetActivity extends Activity {
    EditText edt_fname;
    EditText edt_fusername;
    Button btn_fsure;
```

```java
registerHelper db = new registerHelper(this);
@Override
protected void onCreate(Bundle savedInstanceState) {        // 获取组件
    super.onCreate(savedInstanceState);
    setContentView(R.layout.activity_forget);
    edt_fname = (EditText) findViewById(R.id.edt_fname);
    edt_fusername = (EditText) findViewById(R.id.edt_fusername);
    btn_fsure = (Button) findViewById(R.id.btn_fsure);
    final AlertDialog dialog1 = null ;
    btn_fsure.setOnClickListener(new OnClickListener() {
        @Override
        public void onClick(View arg0) {
                // TODO Auto-generated method stub
            final String str_name = edt_fname.getText().toString().trim();
            final String str_username = edt_fusername.getText().toString().trim();
            // 判断数据库信息与输入信息是否相同
            if(getMsg(str_name,str_username)){        // 两次输入相同
View view=LayoutInflater.from(ForgetActivity.this).inflate(R.layout.updata, null);
final EditText edt_sruepwd=(EditText)view.findViewById(R.id.edt_surepwd);
final EditText edt_sruepwd1 =(EditText) view.findViewById(R.id.edt_surepwd1);
final AlertDialog.Builder builder=new AlertDialog.Builder(ForgetActivity.this);
        AlertDialog dialog ;                // 编写 dialog 密码修改框
        builder.setIcon(android.R.drawable.ic_dialog_info);
        builder.setView(view);
        builder.setPositiveButton(" 确定 ", new DialogInterface.OnClickListener() {
                @Override
            public void onClick(DialogInterface arg0, int arg1) {
                    // TODO Auto-generated method stub
                String str_sruepwd=edt_sruepwd.getText().toString().trim();
                String str_sruepwd1=edt_sruepwd1.getText().toString().trim();
            if(!str_sruepwd.equals(str_sruepwd1)){
        Toast.makeText(ForgetActivity.this, " 两次输入密码不一致 ", 0).show();
                Field field;
        // 点击确定使 dialog 提示框显示
            try {
field = arg0.getClass().getSuperclass().getDeclaredField("mShowing");
        field.setAccessible(true);
        field.set(arg0, false);
```

```java
        } catch (Exception e) {
            // TODO Auto-generated catch block
            e.printStackTrace();}
        return;
            }else{
            db.update(str_sruepwd,str_name,str_username);
            db.close();
            try {
Field field = arg0.getClass().getSuperclass().getDeclaredField("mShowing");
    field.setAccessible(true);
    field.set(arg0, true);
        } catch (Exception e) {
            // TODO Auto-generated catch block
            e.printStackTrace();
        }   }   }        });
builder.setNegativeButton(" 取消 ", new DialogInterface.OnClickListener() {
            @Override
    // 点击取消 dialog 显示框消失
    public void onClick(DialogInterface arg0, int arg1) {
        // TODO Auto-generated method stub
        try {
Field field = arg0.getClass().getSuperclass().getDeclaredField("mShowing");
    field.setAccessible(true);
    field.set(arg0, true);
        } catch (Exception e) {
            // TODO Auto-generated catch block
            e.printStackTrace();
    }   }   });
            AlertDialog dialog1 = builder.create();
            dialog1.show();
        }else{
    // 两次输入不同
    Toast.makeText(ForgetActivity.this, " 验证信息错误 ", 0).show();
        }   }   }); }
    // 调用数据库信息
    private Boolean getMsg(String name, String userpwd) {
        // TODO Auto-generated method stub
        Cursor c = db.equew(name, userpwd);
```

```java
System.out.println(c.getCount() + "-----------------");
if (c.getCount() > 0) {
        return true;
} else {
        return false;
} } }
```

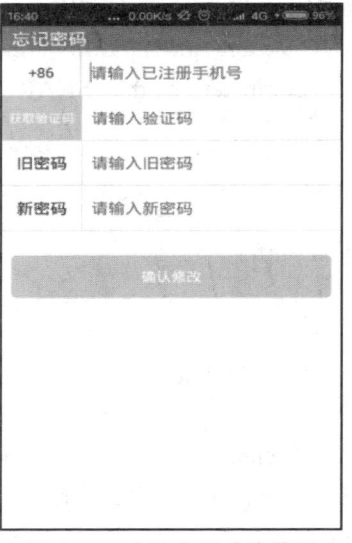

图 3.20　忘记密码功能界面

第十步：通过获得第三方平台（腾讯 QQ）授权，将登录信息上传到服务器，实现 QQ 快捷登录，具体如代码 CORE0311 所示。效果如图 3.21 所示。

代码 CORE0311　QQ 快捷登录

```java
/* 此处添加 QQ 快捷登录的代码 */
private void inItSsoHolder() {
UMQQSsoHandler qqSsoHandler = new UMQQSsoHandler(LoginActivity.this,
"1104587605", "pmaD0FRatx3CoM1R");              // qq 授权
        qqSsoHandler.addToSocialSDK();
        mController.getConfig().setSsoHandler(new SinaSsoHandler());   }
public void onClick(View v) {
        switch (v.getId()) {
        case R.id.btn_qq:
mController.doOauthVerify(LoginActivity.this,  SHARE_MEDIA.QQ, new UMAuth-
Listener() {
        @Override
```

```java
                    public void onStart(SHARE_MEDIA platform) {
                }
                @Override
        public void onError(SocializeException e, SHARE_MEDIA platform) {
        Toast.makeText(LoginActivity.this, " 授权错误 ", Toast.LENGTH_SHORT).show();
                }
                @Override
            public void onComplete(Bundle value, SHARE_MEDIA platform) {
        appContext.setLogin_name(value.get("uid").toString());
        mController.getPlatformInfo(LoginActivity.this, SHARE_MEDIA.QQ, new UMDataL-
istener() {
    // 获取相关授权信息
                @Override
                    public void onStart() {
                                    }
                @Override
                    public void onComplete(int status, Map<String, Object> info) {
                            if (status == 200 && info != null) {
                        appContext.setNickname(info.get("screen_name").toString());
                        appContext.setHead_url(info.get("profile_image_url").toString());
                                login();}
                            else {
                                    Log.d("TestData", " 授权失败 " + status);
                                        } } }); }
                @Override
                public void onCancel(SHARE_MEDIA platform) {
        Toast.makeTextLoginActivity.this, " 授权取消 ", Toast.LENGTH_SHORT).show();
                        } });
                    break;
                    case R.id.btn_sina:
        mController.doOauthVerify(LoginActivity.this, SHARE_MEDIA.SINA, new UMAuth-
Listener() {
        @Override
    public void onError(SocializeException e, SHARE_MEDIA platform) {
                    }
        @Override
    public void onComplete(Bundle value, SHARE_MEDIA platform) {
            if (value != null&& !TextUtils.isEmpty(value.getString("uid"))) {
```

```java
mController.getPlatformInfo(LoginActivity.this, SHARE_MEDIA.SINA, new UMDa-
taListener() {
        @Override
            publicvoid onStart() {
Toast.makeText(LoginActivity.this, " 获取平台数据开始 ...",
Toast.LENGTH_SHORT).show();
}
        @Override
public void onComplete(int status, Map<String, Object> info) {
        if (status == 200 && info != null) {
appContext.setNickname(info.get("nickname").toString());
appContext.setLogin_name(info.get("openid").toString());
appContext.setHead_url(info.get("headimgurl").toString());
    login();
                } else {
                Log.d("TestData", " 授权失败 " + status);
                        } } });
                } else {
Toast.makeText(LoginActivity.this, " 授权失败 ", Toast.LENGTH_SHORT).show();
                        } }
        @Override
        public void onCancel(SHARE_MEDIA platform) {
                }
        @Override
        public void onStart(SHARE_MEDIA platform) {
                } });

            break;
            default:
            break;
            } }
    private void login() {
String url ="http://localhost:8080/Watch/UserThreeLogin?operation=three
Login&nickname=1232&phoneNumber=34567fdgfdbgv&avatorId=http://local;8080/dgdf.jsp";
        String url ="http://123.56.143.204:8080/Watch/UserThreeLogin";
        String operation = "threeLogin";
        String username = appContext.getLogin_name();
        String nickname = appContext.getNickname();
```

```java
        String head_img = appContext.getHead_url();
        RequestParams params = new RequestParams();
        params.add("operation", operation);
        params.add("phoneNumber", username);
        params.add("nickname", nickname);
        params.add("avatorId", head_img);
AsyncHttpHelper.postAbsoluteUrl(url, params, new TextHttpResponseHandler() {
        @Override
        public void onSuccess(int arg0, Header[] arg1, String arg2) {
            try {
                JSONObject obj = new JSONObject(arg2);
                int user_id = obj.optInt("userId");
                appContext.setUser_id(user_id);
Toast.makeText(appContext, user_id+"", Toast.LENGTH_SHORT).show();
                LoginActivity.this.finish();
                setResult(200);
            } catch (JSONException e) {
                e.printStackTrace();
            } }
        @Override
        public void onFailure(int arg0, Header[] arg1, String arg2, Throwable t) {
        } }); }
```

图 3.21　QQ 账号登录界面

第十一步：通过获取第三方平台（新浪微博）授权，实现新浪快捷登录，具体如代码

项目三　登录注册　　　77

CORE0312 所示。效果如图 3.22 所示。

代码 CORE0312　新浪快捷登录

```
/* 此处添加新浪快捷登录代码 */
@Override
    public void onClick(View v) {
            switch (v.getId()) {
case R.id.btn_sina:
mController.doOauthVerify(LoginActivity.this, SHARE_MEDIA.SINA, new UMAuth-
Listener() {
@Override
public void onError(SocializeException e, SHARE_MEDIA platform) {
            }
    @Override
    public void onComplete(Bundle value, SHARE_MEDIA platform) {
    if (value != null && !TextUtils.isEmpty(value.getString("uid"))) {
mController.getPlatformInfo(LoginActivity.this, SHARE_MEDIA.SINA, new UMDa-
taListener() {
    @Override
    public void onStart() {
    Toast.makeText(LoginActivity.this, " 获 取 平 台 数 据 开 始 ...", Toast.LENGTH_
SHORT).show();}
    @Override
    public void onComplete(int status, Map<String, Object> info) {
    if (status == 200 && info != null) {
        appContext.setNickname(info.get("nickname").toString());
        appContext.setLogin_name(info.get("openid").toString());
        appContext.setHead_url(info.get("headimgurl").toString());
        login();
            } else {
                    Log.d("TestData", " 授权失败 " + status);
                            } } });
            } else {
Toast.makeText(LoginActivity.this, " 授权失败 ", Toast.LENGTH_SHORT).show();
                            } }
                    @Override
                    public void onCancel(SHARE_MEDIA platform) {
                    }
```

```
                                @Override
                                public void onStart(SHARE_MEDIA platform) {
                                } });
                break;
            default:
                break;
        } }
```

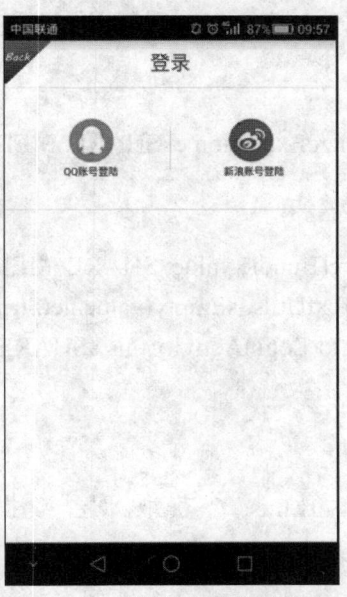

图 3.22　新浪账号登录界面

第十二步：点击"back"按键时，返回到上一界面，具体代码如 CORE0313 所示。效果如图 3.23 所示。

```
代码 CORE0313　返回上一级
/* 此处添加返回上一级代码 */
@Override
    public void onClick(View v) {
            switch (v.getId()) {
case R.id.btn_back:
            LoginActivity.this.finish();
            setResult(200);
            break;
default:
            break;
        } }
```

登录

图 3.23 返回主界面

第十三步：在 SlidingMenu 界面中，点击"系统设置"按键，进入系统设置界面，实现界面初始化，具体如代码 CORE0314 所示。效果如图 3.24 所示。

代码 CORE0314 系统设置

```
@Override
public void onClick(View v) {
        switch (v.getId()) {
// 点击系统设置按键，进入系统设置界面
    case R.id.iv_setting:
    startActivity(new Intent(MainActivity.this,SystemSettingActivity.class));
    break;
default:
    break;
        } }
/* 此处添加系统设置界面初始化功能代码 */
public class SystemSettingActivity extends Activity implements OnClickListener {
    private AppContext appContext;
    private Button btn_back;
    private ImageView iv_txtsize;          // 字体设置
    private ImageView iv_line;             // 离线设置
    private CheckBox cb_push;              // 推送
    private CheckBox cb_wifi;              // WIFI 下加载图片
    private ImageView iv_clear;            // 缓存清理
    private CheckBox cb_day;               // 夜间模式
    private ImageView iv_ret;              // 意见反馈
    private ImageView iv_about;
    private TextView tv_ram;               // 缓存数量
    private Button btn_exit;
    private DiskLruCache TDiskLruCache;
    @Override
    protected void onCreate(Bundle savedInstanceState) {
        super.onCreate(savedInstanceState);
        setContentView(R.layout.activity_sys_setting);
        findById();
```

```
        }
        private void findById() {
                btn_back = (Button) findViewById(R.id.btn_back);
                tv_ram = (TextView) findViewById(R.id.tv_ram);
                iv_clear = (ImageView) findViewById(R.id.iv_clear);
                iv_ret = (ImageView) findViewById(R.id.iv_ret);
                btn_exit = (Button) findViewById(R.id.btn_exit);
                iv_about = (ImageView) findViewById(R.id.iv_about);
                iv_txtsize = (ImageView) findViewById(R.id.iv_txtsize);
                iv_txtsize.setOnClickListener(this);
                iv_about.setOnClickListener(this);
                btn_back.setOnClickListener(this);
                iv_clear.setOnClickListener(this);
                iv_ret.setOnClickListener(this);
                btn_exit.setOnClickListener(this);
        } }
```

图 3.24　添加系统设置

第十四步：实现系统设置中清除缓存、提交意见、系统字体、关于我们等功能，具体如代码 CORE0315 所示。效果如图 3.25 所示。

代码 CORE0315　系统设置
/* 此处添加系统设置功能代码 */ private void InitiCache() {

```java
        try {
    File Bitcache = getDiskCacheDir(SystemSettingActivity.this, "thumb");
    /* 获取图片缓存路径 */
            if (!Bitcache.exists()) {
            Bitcache.mkdirs();
                }
    TDiskLruCache=DiskLruCache.open(Bitcache,getAppVersion(SystemSettingActivity.
this),
    1, 10 * 1024 * 1024);          // 创建 DiskLruCache 实例，初始化缓存数据
        } catch (IOException e) {
        e.printStackTrace();
            } }
    /* 根据传入的 uniqueName 获取硬盘缓存的路径地址。*/
    public File getDiskCacheDir(Context context, String uniqueName) {
        String cachePath;
        if
(Environment.MEDIA_MOUNTED.equals(Environment.getExternalStorageState())
|| !Environment.isExternalStorageRemovable()) {
                cachePath = context.getExternalCacheDir().getPath();
            } else {
                cachePath = context.getCacheDir().getPath();
            }
                returnnew File(cachePath + File.separator + uniqueName);
        }
        /* 获取当前应用程序的版本号。*/
        public int getAppVersion(Context context) {
    try {
    PackageInfo   info   =   context.getPackageManager().getPackageInfo(context.getPacka-
geName(), 0);
        return info.versionCode;
        } catch (NameNotFoundException e) {
            e.printStackTrace();
        } return 1;
        }
        /* 清除所有的缓存 */
        private void DeleCache() {

        try {
```

```java
            TDiskLruCache.delete();
        } catch (IOException e) {
            e.printStackTrace();
        } }
    @Override
    public void onClick(View v) {
        switch (v.getId()) {
        case R.id.btn_back:                    // 返回
            SystemSettingActivity.this.finish();
            setResult(200);
            break;
        case R.id.iv_clear:                    // 清除缓存
            DeleCache();
            tv_ram.setText(" 0 ");
            break;
        case R.id.iv_ret:                      // 意见提交
            strtActivity(new Intent(SystemSettingActivity.this, RetActivity.class));
            break;
        case R.id.btn_exit:
            appContext.setLogin_name("");
            appContext.setNickname("");
            appContext.setHead_url("");
            SystemSettingActivity.this.finish();
            break;
        case R.id.iv_about:                    // 关于我们
            startActivity(new Intent(SystemSettingActivity.this, AboutActivity.class));
            break;
        case R.id.iv_txtsize:                  // 系统字体大小设置
            showDialog();
            break;
        default:
            break;
        } }
    private void showDialog() {
        final String[] arrayFruit = new String[] { " 大号 "," 中号 "," 小号 " };
        finalint[] array = newint[]{ 1, 2, 3 };
Dialog alertDialog = new AlertDialog.Builder(this).setTitle(" 正文字体大小？ ").setI-
con
```

项目三 登录注册   83

```java
(R.drawable.ic_launcher).setItems(arrayFruit, new DialogInterface.OnClickListener() {
    @Override
    public void onClick(DialogInterface dialog, int which) {
        switch (array[which]) {
            case 1:
                appContext.setTextSize(3);
                break;
            case 2:
                appContext.setTextSize(2);
                break;
            case 3:
                appContext.setTextSize(1);
                break;
            default:
                break;
        } }
    }).setNegativeButton(" 取消 ", new DialogInterface.OnClickListener() {
        @Override
        public void onClick(DialogInterface dialog, int which) {
        }
    }).create();
        alertDialog.show();
}
```

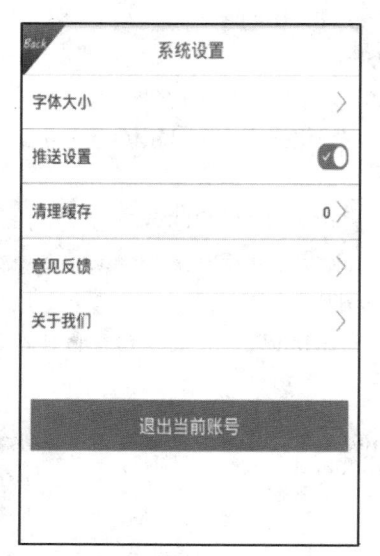

图 3.25   系统设置界面

第十五步：设置分享信息，实现微信、新浪微博、腾讯 QQ 的第三方好友分享功能，具体如代码 CORE0316 所示。效果如图 3.26 所示。

```
代码 CORE0316    第三方好友分享
/* 此处添加好友分享代码 */
private void inItPlatform() {
        addWXPlatform();                        // 添加微信分享平台
        addQQQZonePlatform();               // 添加 QQ 分享平台
        addSinaPlatform();                       // 添加新浪分享平台
        addShareContent();                      // 添加分享
    }
/*@ 功能描述：添加微信平台分享 @return */
private void addWXPlatform() {
    String appId = "wx9d6b6d59cfe13bea";
    String appSecret = "97f64d53a275b1477c4078327f4de6a8";
UMWXHandler wxHandler = new UMWXHandler(MainActivity.this, appId, appSe-
cret);
    wxHandler.addToSocialSDK();              // 添加微信平台
UMWXHandler wxCircleHandler = new UMWXHandler(MainActivity.this, appId,
appSecret);                       // 支持微信朋友圈
        wxCircleHandler.setToCircle(true);
        wxCircleHandler.addToSocialSDK();
    }
/* @ 功能描述：添加 QQ 空间平台分享 @return */
private void addQQQZonePlatform() {
    String appId = "1103838283";
    String appKey = "bAPjZ48G237mKfjR";
UMQQSsoHandler qqSsoHandler = new UMQQSsoHandler(MainActivity.this, appId,
appKey);
        // 添加 QQ 支持，并且设置 QQ 分享内容的 target url
        qqSsoHandler.setTargetUrl("http://www.umeng.com/social");
        qqSsoHandler.addToSocialSDK();
    QZoneSsoHandler qZoneSsoHandler = new QZoneSsoHandler(MainActivity.this, ap-
pId, appKey);
        // 添加 QZone 平台
        qZoneSsoHandler.addToSocialSDK();
    }
    /* @ 功能描述：添加新浪平台分享 @return */
```

```java
    private void addSinaPlatform() {
        SinaSsoHandler sinaSsoHandler = new SinaSsoHandler();
            sinaSsoHandler.addToSocialSDK();

    }
    private void addShareContent() {
            WeiXinShareContent weiXinShareContent = new
WeiXinShareContent();
            weiXinShareContent.setShareContent("U 酒保,您随身必带的小酒保！");
            weiXinShareContent.setTitle("SpiritHelper");
    weiXinShareContent.setShareImage(new    UMImage(MainActivity.this,    R.drawable.
icon));
            weiXinShareContent.setTargetUrl("http://www.umeng.com/social");
            mController.setShareMedia(weiXinShareContent);
    QQShareContent qqShareContent = new QQShareContent();
            qqShareContent.setShareContent("U 酒保,您随身必带的小酒保！");
            qqShareContent.setTitle("SpiritHelper");
    qqShareContent.setShareImage(new UMImage(MainActivity.this, R.drawable.icon));
            qqShareContent.setTargetUrl("http://www.umeng.com/social");
            mController.setShareMedia(qqShareContent);
            QZoneShareContent qZoneShareContent = new QZoneShareContent();
            qZoneShareContent.setShareContent("U 酒保,您随身必带的小酒保！");
            qZoneShareContent.setTitle("SpiritHelper");
    qZoneShareContent.setShareImage(new    UMImage(MainActivity.this,    R.drawable.
icon));
            qZoneShareContent.setTargetUrl("http://www.umeng.com/social");
            mController.setShareMedia(qZoneShareContent);
            SinaShareContent SinaShareContent = new SinaShareContent();
            SinaShareContent.setShareContent("U 酒保,您随身必带的小酒保！");
            SinaShareContent.setTitle("SpiritHelper");
    SinaShareContent.setShareImage(new UMImage(MainActivity.this, R.drawable.icon));
            SinaShareContent.setTargetUrl("http://www.umeng.com/social");
            mController.setShareMedia(SinaShareContent);

    }
```

图 3.26　第三方好友分享

第十六步：运行项目，实现如图 3.7 至图 3.12 所示效果。

任务总结

本项目介绍了 U 酒保登录模块功能的实现，通过对本项目的学习可以了解手机侧滑菜单、注册登录的实现方法以及 ShareSDK 分享机制，重点掌握了 SlidingMenu 使用方法，实现侧滑菜单效果以及登录注册功能的方法。

英语角

sliding	滑行的	menu	菜单
mode	方式	offset	抵消
script	脚本	login	登录
info	信息	touch	触摸
margin	边缘	enabled	激活的

任务习题

一、选择题

1. 在手机开发中常用的数据库是（　　　）。

A.SQLite　　　　　　B.Oracle　　　　　　C.SQL Server　　　　　　D.Db23

2. 关于 SQLite 数据库的操作，下面说法不正确的是（　　　）。

A.Context 对象调用 openOrCreateDatabase() 方法打开或者创建数据库

B.SQLiteDatabase 类的静态方法 openOrCreateDatabase() 方法打开或者创建数据库

C.Context 对象 closeDatabase() 方法关闭数据库

D.SQLiteDatabase 类的静态方法 deleteDatabase() 方法删除数据库

3. 如果在 Android 应用程序中需要发送短信,那么需要在 AndroidManifest.xml 文件中增加什么样的权限（　　）。

A. 发送短信,无需配置权限 　　　　B.permission.SMS

C.android.permission.RECEIVE_SMS 　　D.android.permission.SEND_SMS

4.SlidingMenu 设置阴影图片,通过以下那种属性实现（　　）。

A.menu.setMode(SlidingMenu.LEFT)

B.menu.setShadowDrawable(R.drawable.shadow)

C.menu.setBehindOffsetRes(R.dimen.slidingmenu_offset)

D.menu.setShadowWidthRes(R.dimen.shadow_width)

5. 本项目中用到了第三方 SDK 实现分享功能和短信验证功能,用到的 SDK 分别为（　　）和短信验证 SDK。

A.ShareSDK　　　　B.BBSSDK　　　　C.ShareREC　　　D.MobAPI

二、填空题

1.SlidingMenu 的是一种比较新颖的 _____ 或 _____ 的技术。

2. 手机应用为提高其 _____ ,通过使用服务器向用户发送验证码的方式,保护用户个人信息安全。

3.ShareSDK for Android 版正式发布日期为 _____ 。

4.ShareSDK 按优先级分为三种分别是 _____ 、_____ 、_____ 。

5. 手机短信验证是一种常见的功能,通过 _____ 实现该功能。

三、上机题

1. 实现左右侧滑功能。

2. 实现短信验证功能。

项目四 酒精检测

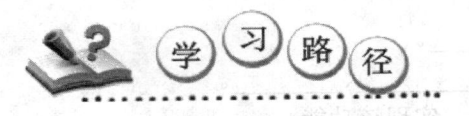

通过 U 酒保项目酒精检测模块的实现,了解蓝牙连接方式,掌握复杂 JSON 解析方法,学习使用进度条显示酒精浓度,具有使用进度条显示数据的能力。在任务实现过程中:

● 了解蓝牙匹配形式。
● 了解蓝牙通信协议及方法。
● 掌握 JSON 多种解析方式。

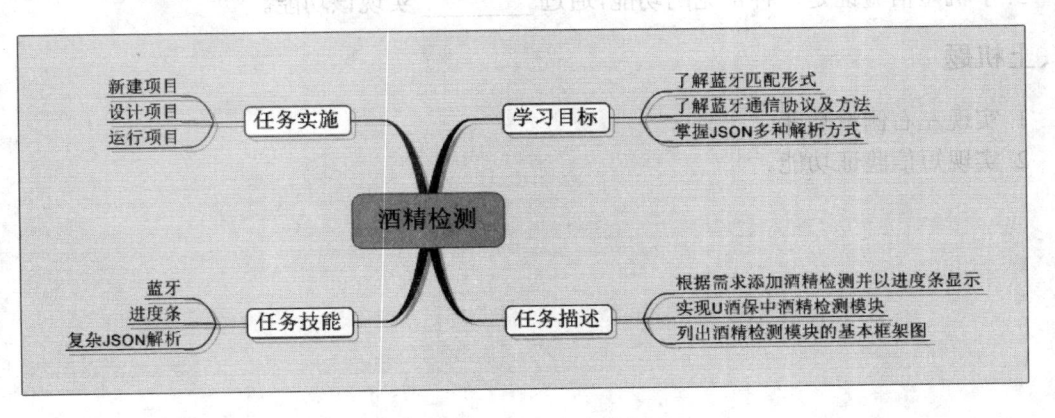

【情境导入】

酒精浓度的测试方法有很多,常见的是交警手中的酒精检测仪,通过驾驶员呼出气体进行酒精含量的检测。U 酒保软件的开发能够使酒精检测更加便捷。将手机与检测装置通过蓝牙连接后,进入主界面,点击屏幕上的"立即检测"的同时向检测仪呼气,在界面显示相应的酒精浓度值。本项目通过酒精检测模块的实现,讲解蓝牙通信以及圆形进度条的使用方法。

项目四　酒精检测　　89

【功能描述】

本项目将实现 U 酒保中酒精检测模块：

- 通过相对布局创建进度条界面。
- 调用自定义组件实现圆形进度条。
- 创建蓝牙连接，连接到酒精检测仪，并接收检测仪返回的酒精浓度。
- 通过进度条百分比显示酒精浓度。

【基本框架】

基本框架如图 4.1 所示。通过本模块的学习，能将框架图 4.1 转换成效果图 4.2 所示。

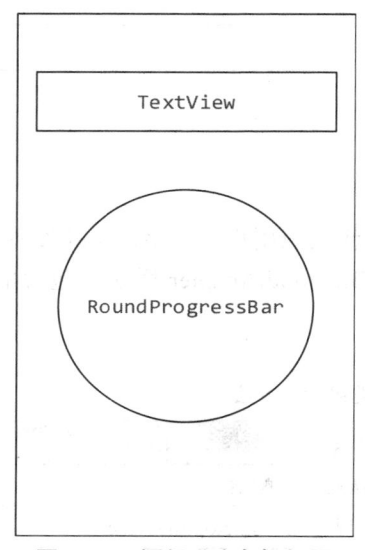

图 4.1　U 酒保进度条框架图

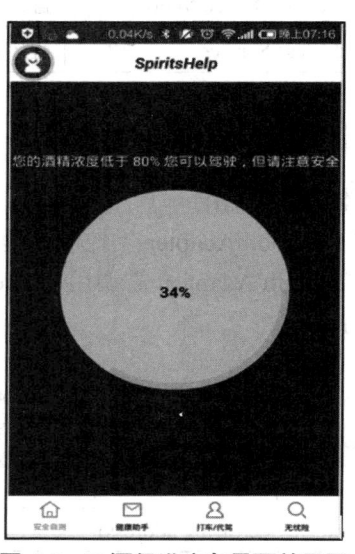

图 4.2　U 酒保进度条界面效果图

技能点 1　蓝牙

1　蓝牙简介

　　蓝牙（Bluetooth）是目前使用最广泛的无线通信协议，主要针对短距离的设备通讯，可实现移动设备和楼宇（或个人）域网之间的数据交换，常用于连接耳机，鼠标等。在应用中，蓝牙相关的 API 有两个，BluetoothAdapter 代表本地的蓝牙适配器，BluetoothDevice 代表远程的蓝牙适配器。该两种 API 都属于 android：bluetooth.* 中的类，该类中重要类作用如表 4.1 所示。

Android 模块化项目实战

表 4.1   android.bluetooth.* 中的重要类的作用

类　　名	作　　用
BluetoothAdapter	本地蓝牙设备的适配类,所有的蓝牙操作都要通过该类完成
BluetoothClass	用于描述远端设备的类型,特点等信息
BluetoothDevice	蓝牙设备类,代表了蓝牙通讯过程中的远端设备
BluetoothServerSocket	蓝牙设备服务端,类似 ServerSocket
BluetoothSocket	蓝牙设备客户端,类似 Socket
BluetoothClass.Device	蓝牙关于设备信息
BluetoothClass.Device.Major	蓝牙设备管理
BluetoothClass.Service	蓝牙相关服务

2　蓝牙适配

（1）本地蓝牙适配

通过 BluetoothAdapter 类控制本地蓝牙设备。该类代表应用程序的 Android 设备,为访问默认的 Bluetooth Adapter,需调用 getDefaultAdapter()。BluetoothAdapter 适配类包含的方法如表 4.2 所示。

表 4.2   BluetoothAdapter 适配类包含的方法

方　　法	作　　用
getDefaultAdapter(context)	得到默认蓝牙适配器
getRemoteDevice(String address)	得到指定蓝牙的 BluetoothDevice
isEnabled(int)	蓝牙是否开启
getState(object)	得到蓝牙状态
Enable(boolean)	打开蓝牙
Disable(boolean)	关闭蓝牙
getAddress(byte)	得到蓝牙适配器地址
getName(String)	得到蓝牙的名字
setName(String)	设置蓝牙的名字
getScanMode(builder)	得到当前蓝牙的扫描模式
setScanMode(builder)	设置当前蓝牙的设置模式
startDiscovery(boolean)	开始搜索蓝牙设备
cancelDiscovery(boolean)	取消搜索蓝牙设备
isDiscovering(boolean)	是否允许被搜索
getBondedDevices(Set<BluetoothDevice>)	得到 BluetoothDevice 集合到本地适配器
listenUsingRfcommWithServiceRecord()	创建一个监听,安全记录蓝牙
checkBluetoothAddress(boolean)	检查蓝牙地址是否正确

（2）远程蓝牙适配

BluetoothDevice 对象代表远程蓝牙设备，通过该类可查询远程设备物理地址、连接状态、名称等信息，该类的操作执行在远程蓝牙设备硬件上，对象获取途径：

● 调用 BluetoothAdapter 的 getRemoteDevice() 方法获取该类对象对应的物理地址。

● 调用 BluetoothAdapter 的 getBoundedDevices() 方法，可获取已配对的蓝牙设备集合。

（3）蓝牙的启动方式

蓝牙适配器指蓝牙设备的转换接口，采用了全球通用的短距无线连接技术，读取任何一种本地的 Bluetooth Adapter 属性、启动蓝牙进行扫描、找到绑定设备、修改本地属性，需在 AndroidManifest.xml 文件下添加 BLUETOOTH_ADMIN 使用权限，具体代码如下所示。

```
<uses-permission android:name="android.permission.BLUETOOTH" />
<uses-permission android:name="android.permission.BLUETOOTH_ADMIN" />
```

打开蓝牙的方式有两种：

● Android 系统弹出提示框，提示用户是否开启蓝牙设备，具体代码如下所示。

```
Intent enabler = new Intent(BluetoothAdapter.ACTION_REQUEST_ENABLE);
startActivityForResult(enabler, REQUEST_ENABLE);
```

● 直接打开蓝牙设备，在运用当中视情况而定，有些情况需要提示，有些情况则可直接打开使用，具体代码如下所示。

```
BluetoothAdapter _bluetooth = BluetoothAdapter.getDefaultAdapter();
bluetooth.enable();
```

拓展 在智能手机还没有普及的时候，蓝牙估计是我们使用最早的一种短距离数据传送工具。上边已经对蓝牙的使用介绍的很详细了。那么蓝牙的由来以及发展历史想必没有几个人能细细道来。下边小小的二维码就会为你解开疑惑。

技能点 2　进度条

1　ProgressBar 简介

进度条（ProgressBar）是用于显示某个耗时操作完成百分比的组件，在 UI 界面中非常实用,能够动态地显示操作进度。在执行耗时操作时,用户通过进度条变化确认程序是否失去响应,以此提高用户界面友好性,提高用户体验性。在应用中,系统提供了两大类进度条样式,长形进度条(progressBarStyleHorizontal)和圆形进度条(progressBarStyleLarge)。本项目主要介绍自定义圆形进度条的使用。

2 ProgressBar 属性方法

在两大类进度条样式下又可以分为六种不同风格的进度条,通过设置 ProgressBar 的 style 属性,可实现不同样式的进度条,style 属性设置如表 4.3 所示。

表 4.3　ProgressBar 的 style 属性

属　　性	含　　义
style="@android:style/Widget.ProgressBar.Small"	小型圆形进度条
style="@android:style/Widget.ProgressBar.Small.Inverse"	小型圆形进度条
style="@android:style/Widget.ProgressBar.Inverse"	中型圆形进度条
style="@android:style/Widget.ProgressBar.Large"	大型圆形进度条
style="@android:style/Widget.ProgressBar.Large.Inverse"	大型圆形进度条
style="@android:style/Widget.ProgressBar.Horizontal"	水平进度条

ProgressBar 在运用过程中还支持多种常用 XML 属性如表 4.4 所示。

表 4.4　ProgressBar 支持多种常用 XML 属性

XML 属性	说　　明
android:max	设置该进度条的最大值
android:progress	设置该进度条的已完成进度值
android:progressDrawable	设置该进度条的轨道对应的 Drawable 对象
android:indeterminate	该属性设置为 true,设置进度条不精确显示进度
android: indeterminateDrawable	设置绘制不显示进度的进度条的 Drawable 对象
android: indeterminateDuration	设置不精确显示进度的持续时间
android: progressBarStyle	默认进度条样式
android: progressBarStyleHorizontal	水平进度条样式
android: progressBarStyleLarge	大进度条样式
android: progressBarStyleSmall	小进度条样式

在 Java 代码中,ProgressBar 提供了以下方法来操作进度:

● setProgress(int): 设置进度条的完成百分比。

● incrementProgressBy(int): 设置进度条的进度。当 int 为正数时进度增加;为负数时进度减少。

项目四　酒精检测　　93

● tileify(drawable, false) 方法和 tileifyIndeterminate(drawable) 方法。这两个方法主要是对 Drawable 进行解析、转换的过程。

3　圆形进度条的实现

圆形进度条需要自定义一个 RoundProgressBar 类，自定义 RoundProgressBar 类需要继承 View 类，并添加类的构造方法。通过重写 onDraw () 方法和自定义属性实现圆形进度条，效果如图 4.3 所示。

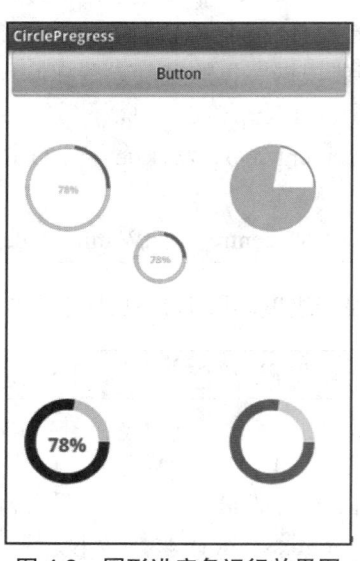

图 4.3　圆形进度条运行效果图

（1）使用 onDraw() 方法画外层的圆环，并设置圆心、半径、颜色等。具体代码如 CORE0401 所示。

```
代码 CORE0401  使用 onDraw() 方法画圆

protected void onDraw(Canvas canvas) {
    super.onDraw(canvas);
    /* 画最外层的大圆环 */
    int centre = getWidth()/2;              // 获取圆心的 x 坐标
    int radius = (int) (centre - roundWidth/2);    // 圆环的半径
    paint.setColor(roundColor);             // 设置圆环的颜色
    paint.setStyle(Paint.Style.STROKE);     // 设置空心
    paint.setStrokeWidth(roundWidth);       // 设置圆环的宽度
    paint.setAntiAlias(true);               // 消除锯齿
    canvas.drawCircle(centre, centre, radius, paint);  // 画出圆环
```

（2）设置进度百分比以及字体宽度。具体代码如 CORE0402 所示。

代码 CORE0402 设置进度百分比以及字体

```
/* 画进度百分比 */
paint.setStrokeWidth(0);                    // 进度从 0 开始
    paint.setColor(textColor);                  // 设置进度条颜色
    paint.setTextSize(textSize);                // 设置字体大小
    paint.setTypeface(Typeface.DEFAULT_BOLD);   // 设置字体
// 中间的进度百分比，先转换成 float 在进行除法运算，不然都为 0
    int percent = (int)(((float)progress / (float)max) * 100);
// 测量字体宽度，我们需要根据字体的宽度设置在圆环中间
 floattextWidth=paint.measureText(percent+"%");
 if(textIsDisplayable && percent != 0 && style == STROKE){
// 画出进度百分比
canvas.drawText(percent + "%", centre - textWidth / 2, centre + textSize/2, paint);   }
```

（3）设计圆弧圆环样式，并根据进度画圆弧。具体代码如 CORE0403 所示。

代码 CORE0403 设计圆弧圆环进度

```
/* 画圆弧，画圆环的进度 */
    // 设置进度是实心还是空心
    paint.setStrokeWidth(roundWidth);           // 设置圆环的宽度
    paint.setColor(roundProgressColor);         // 设置进度的颜色
RectF oval = new RectF(centre - radius, centre - radius, centre + radius, centre + radi-
us);
    // 用于定义的圆弧的形状和大小的界限
    switch (style) {
    case STROKE:{
      paint.setStyle(Paint.Style.STROKE);       // 根据进度画圆弧
     canvas.drawArc(oval, 0, 360 * progress / max, false, paint);
      break;
     }
    case FILL:{
      paint.setStyle(Paint.Style.FILL_AND_STROKE);
      if(progress !=0) {
 canvas.drawArc(oval, 0, 360 * progress / max, true, paint);  // 根据进度画圆弧
      break;
    } }    }
```

项目四　酒精检测

技能点 3　复杂 JSON 解析

1　JSON 简介

　　JSON（JavaScript Object Notation）是一种轻量级的数据交换格式，基于 ECMAScript 的一个子集。JSON 采用独立于语言的文本格式，但也使用了类似于 C 语言家族的习惯（包括 C、C++、C#、Java、JavaScript、Perl、Python 等）。易于阅读、编写、解析和生成（一般用于提升网络传输速率）。其中 JSON 格式如表 4.5 所示。

表 4.5　JSON 格式

类　型	数据结构	含　义
字符串	{key: value,key: value,...}	key: 对象的属性；value: 对应属性值
数组	["java","javascript","vb",...]	字段值可以是任何类型

2　JSON 解析方法

　　Android 平台自带了 JSON 解析的相关 API，可以将文件、输入流中的数据转化为 JSON 对象，然后从对象中获取 JSON 保存的数据内容。JSON 解析在包 org.json 下，主要有以下几个类：

　　（1）JSONObject

　　可看作是 JSON 对象，是系统中有关 JSON 定义的基本单元，其包含 Key 和 Value 数值。它对外部调用的响应体现为一个标准的字符串（例如：{"JSON": "Hello, World"}，最外层被大括号包裹，其中的 Key 和 Value 用冒号分隔。

　　（2）JSONStringer

　　JSON 文本构建类（JSONStringer），该类可以帮助快速便捷的创建 JSON text。其最大的优点在于可以减少格式错误导致的程序异常，引用这个类可以自动严格按照 JSON 语法规则（syntax rules）创建 JSON text。每个 JSONStringer 实体只能对应创建一个 JSON text。

　　（3）JSONArray

　　代表一组有序数值。将其转换为 String 输出，其形式使用方括号包裹，数值以逗号分隔（例如：[value1,value2,value3]），利用简短的代码更加直观的了解其格式。该类的内部具有查询功能，get() 和 put () 两种方法都可以通过 index 索引返回指定的数值，put() 方法用来添加或者替换数值。该类的 value 类型包括：Boolean、JSONArray、JSONObject、Number、String 或者默认值 JSONObject.NULL object。

　　（4）JSONTokener

　　JSON 解析类。

（5）JSONException

JSON.org 类抛出的异常信息。

3　JSON 解析实现

（1）传统 JSON 解析

第一步：生成 JSON 字符串，具体代码如下所示。

```
public static String createJsonString(String key, Object value) {
JSONObject jsonObject = new JSONObject();          // 遍历 JSONObject
jsonObject.put(key, value);                        // 在 JSONObject 中填充内容
return jsonObject.toString();                      // 返回 JSON 字符串
}
```

第二步：解析 JavaBean、List 数组、嵌套 Map 的 List 数组 JSON 字符串，具体代码如下所示。

```
Public class JsonTools {
Public static Person getPerson(String key, String jsonString) {
Person person = new Person();
try{
JSONObject jsonObject = new JSONObject(jsonString);    // 初始化 JSONObject 方法
JSONObject personObject = jsonObject.getJSONObject("person");
person.setId(personObject.getInt("id"));               // 添加 ID
person.setName(personObject.getString("name"));        // 添加 name
person.setAddress(personObject.getString("address"));  // 添加 address
} catch(Exception e) {
// TODO: handle exception
}
return person;
}
Public staticList<person> getPersons(String key, String jsonString) {
List<person> list = newArrayList<person>();
try{
JSONObject jsonObject = newJSONObject(jsonString);
JSONArray jsonArray = jsonObject.getJSONArray(key);    // 返回 JSON 的数组
for(int i = 0; i < jsonArray.length(); i++) {          // 遍历得到 JSon 数组中的数
JSONObject jsonObject2 = jsonArray.getJSONObject(i);
Person person = new Person();
person.setId(jsonObject2.getInt("id"));
```

项目四　酒精检测　　　　　　　　　　　　　　　　　　　　　　97

```java
person.setName(jsonObject2.getString("name"));
person.setAddress(jsonObject2.getString("address"));
list.add(person);
}
} catch(Exception e) {
// TODO: handle exception
}
return list;
}
Public static List<string> getList(String key, String jsonString) {// 得到 JSON 加载到
List
List<string> list = newArrayList<string>();
try{
JSONObject jsonObject = new JSONObject(jsonString);
JSONArray jsonArray = jsonObject.getJSONArray(key);
for(int i = 0; i < jsonArray.length(); i++) {
String msg = jsonArray.getString(i);
list.add(msg);
}
} catch(Exception e) {
// TODO: handle exception
}
return list;
}
Public static List<map<string, object="">> list KeyMaps(String key, String json-
String) {
// 从 List 中嵌套 Map 读取值
    List<map<string, object="">> list = new ArrayList<map<string, object="">>();
    try{
        JSONObject jsonObject = newJSONObject(jsonString);
        JSONArray jsonArray = jsonObject.getJSONArray(key);
        for(inti = 0; i < jsonArray.length(); i++) {
            JSONObject jsonObject2 = jsonArray.getJSONObject(i);
            Map<string, object=""> map = new HashMap<string, object="">();
            Iterator<string> iterator = jsonObject2.keys();
            while(iterator.hasNext()) {
                String json_key = iterator.next();
                Object json_value = jsonObject2.get(json_key);
```

```
            if(json_value == null) {
                json_value = "";
            } map.put(json_key, json_value);
        } list.add(map);
    } } catch(Exception e) {
    // TODO: handle exception
} return list;
} }
```

（2）JSON 解析之 GSON

第一步：生成 JSON 字符串，具体代码如下所示。

```
public class JsonUtils {
public static String createJsonObject(Object obj) {
  Gson gson = new Gson();
  String str = gson.toJson(obj);
  return str;
} }
```

第二步：解析 JSON 串，具体代码如下所示。

```
public class GsonTools {
    public GsonTools() {
        // TODO Auto-generated constructor stub
    }
    public static <t> T getPerson(String jsonString, Class<t> cls) {
        T t = null;
        try {
            Gson gson = new Gson();
            t = gson.fromJson(jsonString, cls);
        } catch (Exception e) {
        // TODO: handle exception
        } return t;
    }
    /* 使用 GSON 进行解析 List<person> */
    public static <t> List<t> getPersons(String jsonString, Class<t> cls) {
        List<t> list = new ArrayList<t>();
        try {
            Gson gson = new Gson();
```

项目四 酒精检测　　　　99

```java
            list = gson.fromJson(jsonString, new TypeToken<list<t>>() {
            }.getType());
        } catch (Exception e) {
        }return list;
    }
    public static List<string> getList(String jsonString) {
        List<string> list = new ArrayList<string>();
        try {
            Gson gson = new Gson();
            list = gson.fromJson(jsonString, new TypeToken<list<string>>() {
            }.getType());
        } catch (Exception e) {
            // TODO: handle exception
        }return list;
    }
    public static List<map<string, object="">> listKeyMaps(String jsonString) {
        List<map<string, object="">> list = new ArrayList<map<string, object="">>();
        try {
            Gson gson = new Gson();
            list = gson.fromJson(jsonString, new TypeToken<list<map<string, object="">>>(){
            }.getType());
        } catch (Exception e) {
        // TODO: handle exception
        }return list;
    }  }
```

（3）JSON 解析之 FastJSON，具体代码如下所示。

```java
public class JsonTool {
public static <t> T getPerson(String jsonstring, Class<t> cls) {
T t = null;
try {
t = JSON.parseObject(jsonstring, cls);
} catch (Exception e) {
// TODO: handle exception
}return t;
}
// 解析 FastJSON
```

```
public static <t> List<t> getPersonList(String jsonstring, Class<t> cls) {
List<t> list = new ArrayList<t>();
try {
list = JSON.parseArray(jsonstring, cls);
} catch (Exception e) {
// TODO: handle exception
}return list;
}
public static <t> List<map<string, object="">> getPersonListMap1(
String jsonstring) {
List<map<string, object="">> list = new ArrayList<map<string, object="">>();
try {
list = JSON.parseObject(jsonstring,
new TypeReference<list<map<string, object="">>>() {
}.getType());
} catch (Exception e) {
// TODO: handle exception
}return list;
}  }
```

拓展 通过学习我们已经掌握 JSON 解析的方法及原理。那么有没有想过 JSON 是怎么由来的？其实作者早已为大家整理出来了，顺便列出来 JSON 与 XML 两者之间的区别以及各自的优势。扫描下方二维码即可一探究竟。

通过如下步骤实现 U 酒保的酒精检测模块。

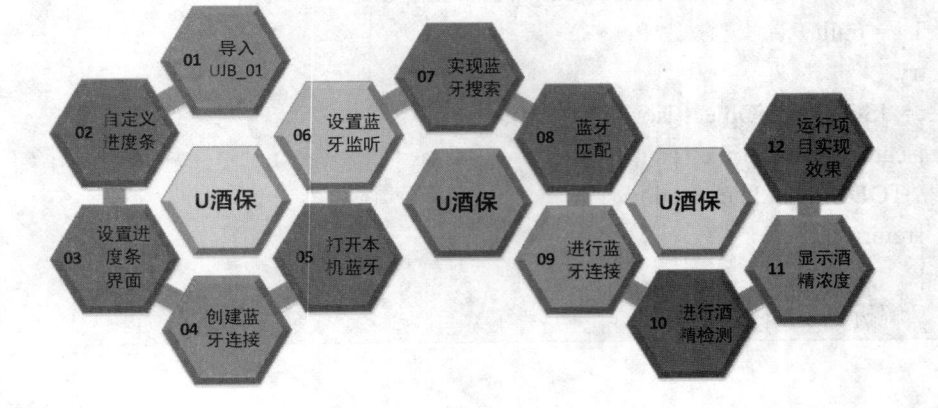

具体步骤如下所示。

第一步：将 UJB_01 导入工程，在其基础上进一步实现 UJB 项目酒精检测模块。首先点击"Open an existing Android Studio project"打开磁盘路径查找所需项目并导入，如图 4.4 和图 4.5 所示。实现如图 4.6 所示结果图。

图 4.4　导入项目

图 4.5　工程目录

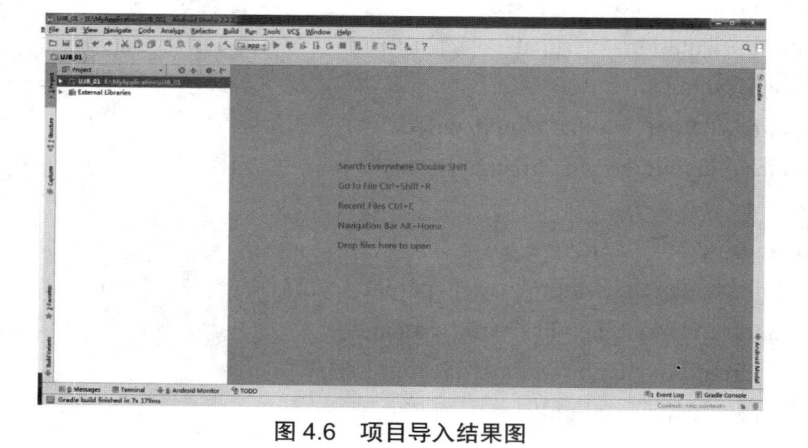

图 4.6　项目导入结果图

第二步：新建 RoundProgressBar.java 实现自定义进度条编写。具体如代码 CORE0404 所示。

代码 CORE0404 自定义进度条代码

```java
/* 此处添加自定义进度条代码 */
@Override
    protected void onDraw(Canvas canvas) {
            super.onDraw(canvas);
            /* 画最外层的大圆 */
            int centre = getWidth()/2;                  // 获取圆心的 x 坐标
            int radius = (int) (centre - roundWidth/2);      // 圆环的半径
            paint.setColor(roundColor);                 // 设置圆环的颜色
            paint.setStyle(Paint.Style.FILL_AND_STROKE);  // 设置空心
            paint.setStrokeWidth(roundWidth);            // 设置圆环的宽度
            paint.setAntiAlias(true);                    // 消除锯齿
            canvas.drawCircle(centre, centre, radius, paint);  // 画出圆环
                    }
```

第三步：调用 RoundProgressBar.java 显示进度条界面 fragment_home.xml。具体如代码 CORE0405 所示。效果如图 4.7 所示。

代码 CORE0405 进度条界面

```xml
/* 此处添加进度条界面代码 */
<?xml version="1.0" encoding="utf-8"?>
<LinearLayout xmlns:android="http://schemas.android.com/apk/res/android"
    xmlns:app="http://schemas.android.com/apk/res/com.example.lxt.ujb_test"
    android:layout_width="fill_parent"
    android:layout_height="fill_parent"
    android:background="#303030"
    android:orientation="vertical" >
<RelativeLayout
        android:layout_width="match_parent"
        android:layout_height="match_parent" >
<TextView
        android:id="@+id/tv_tip"
        android:layout_width="match_parent"
        android:layout_height="wrap_content"
        android:layout_above="@+id/ProgressBar"
        android:layout_alignParentLeft="true"
```

项目四　酒精检测　　　103

```
        android:layout_marginBottom="26dp"
        android:textSize="16sp"
        android:gravity="center"
        android:text=""
        android:textColor="#FFFFFF" />
<com.example.lxt.ujb_test.view.RoundProgressBar
        android:id="@+id/ProgressBar"
        android:layout_width="250dp"
        android:layout_height="250dp"
        android:layout_centerHorizontal="true"
        android:layout_centerVertical="true"
        android:clickable="true"
        android:visibility="visible"
        app:max="250"
        app:roundColor="#D1D1D1"
        app:roundProgressColor="@android:color/holo_blue_bright"
        app:roundWidth="8dip"
        app:textColor="#000000"
        app:textSize="18sp" />
</RelativeLayout>  </LinearLayout>
```

图 4.7　进度条功能界面

第四步：创建蓝牙服务 BluetoothCommService.java，开启蓝牙监听等待连接，显示连接的状态。具体如代码 CORE0406 所示。

代码 CORE0406 创建蓝牙连接

```
/* 此处添加创建蓝牙服务代码 */
// 创建蓝牙服务
public class BluetoothCommService {
    private synchronized void setState(int state) {
    // 将新状态交给处理程序，以便 UI 活动可以更新
    mState = state;
    mHandler.obtainMessage(HomeFragment.MESSAGE_STATE_CHANGE,    state,
-1).sendToTarget();
    }
    public synchronized int getState() {
        return mState;
    }
    /* 开启监听线 */
    public synchronized void start() {
        if (D)
            Log.d(TAG, "start");
        if (mConnectThread != null) {
        // 清除运行中的线程 mConnectThread 使其为空
        mConnectThread.cancel();
        mConnectThread = null;
        }
        if (mConnectedThread != null) {
        // 清除运行中的线程 mConnectedThread 使其为空
        mConnectedThread.cancel();
        mConnectedThread = null;
        }
    if (mAcceptThread == null) {
        mAcceptThread = new AcceptThread();
        mAcceptThread.start();              // 开启监听线程
        }
        setState(STATE_LISTEN);
    }
    /* 连接一个蓝牙对象首先得清空所有已连接的对象 */
    public synchronized void connect(BluetoothDevice device) {
    if (mState == STATE_CONNECTING) {     // 正在连接状态
```

```java
            if (mConnectThread != null) {
                // 清除运行中的线程 mConnectThread 使其为空
                mConnectThread.cancel();
                mConnectThread = null;
            } }
                // 取消当前运行连接的任何线程
        if (mConnectedThread != null)
                // 清除运行中的线程 mConnectedThread 使其为空
                mConnectedThread.cancel();
                mConnectedThread = null;}
    mConnectThread = new ConnectThread(device); // 开始启动一个线程去连接
        mConnectThread.start();
    setState(STATE_CONNECTING);            // 开启新的连接请求线程
        }
        /* 启动线程 */
        public synchronized void connected(BluetoothSocket socket, BluetoothDevice de-
vice) {
            if (D)
                Log.d(TAG, "connected");
            if (mConnectThread != null) {       // 取消完成连接的线程
    // 清除运行中的线程 mConnectThread 使其为空
    mConnectThread.cancel();
                mConnectThread = null;
            }
            if (mConnectedThread != null) {       // 取消完成连接的线程
                mConnectedThread.cancel();
                mConnectedThread = null;
            }if (mAcceptThread != null) {
                mAcceptThread.cancel();
                mAcceptThread = null;
            }
            // 启动线程管理连接并执行传输
            mConnectedThread = new ConnectedThread(socket);
            mConnectedThread.start();         // 和客户端开始通信
    Message msg = mHandler.obtainMessage
    (HomeFragment.MESSAGE_DEVICE_NAME);
// 将连接的设备的名称返回到 UI 活动
        Bundle bundle = new Bundle();
```

```java
        bundle.putString(HomeFragment.DEVICE_NAME, device.getName());
        msg.setData(bundle);
        mHandler.sendMessage(msg);
        setState(STATE_CONNECTED);
}
public synchronized void stop() {        // 停止方法
        if (D)
            Log.d(TAG, "stop");
        if (mConnectThread != null) {
            mConnectThread.cancel();
            mConnectThread = null;
        }
        if (mConnectedThread != null) {
            mConnectedThread.cancel();
            mConnectedThread = null;
        }
        if (mAcceptThread != null) {
            mAcceptThread.cancel();
            mAcceptThread = null;
        }

            SetState(STATE_NONE);
    }
    public void write(byte[] out) {                    // 创建临时对象
        ConnectedThread r;              // 同步复制的 connectedthread
        synchronized (this) {
            if (mState != STATE_CONNECTED)
                return;
                r = mConnectedThread;   // 得到连接线程
        }r.write(out);                  // 同步进行
    }
    /* 连接失败 */
    private void connectionFailed() {
        setState(STATE_LISTEN);         // 将失败消息发送回活动
        Message msg = mHandler.obtainMessage
(HomeFragment.MESSAGE_TOAST);
        Bundle bundle = new Bundle();
        bundle.putString(HomeFragment.TOAST, "Unable to connect device");
        msg.setData(bundle);
```

项目四　酒精检测　　107

```
                mHandler.sendMessage(msg);
        }
        /* 连接丢失 */
        private void connectionLost() {
            setState(STATE_LISTEN);              // 将失败消息发送回活动
            Message msg = mHandler.obtainMessage
(HomeFragment.MESSAGE_TOAST);
            Bundle bundle = new Bundle();
        bundle.putString(HomeFragment.TOAST, "Device connection was lost");
            msg.setData(bundle);
            mHandler.sendMessage(msg);
        }
        /* 接收线程类似于蓝牙模块端一直等待用户的连接 */
private class AcceptThread extends Thread {
private final BluetoothServerSocket mmServerSocket; //Sokcet 连接本地服务器
        public AcceptThread() {
            BluetoothServerSocket tmp = null;
            try {    // 创建一个新的监听服务器套接字
tmp = mAdapter.listenUsingRfcommWithServiceRecord(NAME, SPP_UUID);
            } catch (IOException e) {
            } mmServerSocket = tmp; // 得到 BluetoothServerSocket 对象
        }
    public void run() {
        setName("AcceptThread");              // 设置线程名称
        BluetoothSocket socket = null;        // 建立一个服务器
        // 如果没有连接,请收听服务器套接字
        while (mState != STATE_CONNECTED) {    // 若没有连接,一直执行
            try {
/* 这是一个阻塞调用,只会返回一个成功的连接或异常它是一个阻塞线程,因此不
要放在 Activity 中 */
                socket = mmServerSocket.accept();
            } catch (IOException e) {
                break;
            }
            if (socket != null) {             // 如果连接被接受
                synchronized (BluetoothCommService.this) {
                switch (mState) {
                case STATE_LISTEN:
```

```
                case STATE_CONNECTING: // 情况正常，启动连接线程
                    connected(socket, socket.getRemoteDevice());
                break;
                case STATE_NONE:
                break;
                case STATE_CONNECTED:
                // 要么没有准备好要么已经连接。终止新的 Socket.
            try {
                    socket.close();
                } catch (IOException e) {
                } break;
                } } } }
            if (D)
                Log.i(TAG, "END mAcceptThread");
        }
            public void cancel() {
                if (D)
Log.d(TAG, "cancel " + this);
                try {
                    mmServerSocket.close();
                } catch (IOException e) {
                } } } }
```

第五步：打开本机蓝牙功能。具体如代码 CORE0407 所示。效果如图 4.8 所示。

代码 CORE0407 打开蓝牙

```
/* 此处添加打开代码 */
    private void OpenBT() {                          // 打开蓝牙
        bluetoothAdapter = BluetoothAdapter.getDefaultAdapter();   // 获取到实例
        if (!bluetoothAdapter.isEnabled()) {
Intent enableIntent = new Intent(BluetoothAdapter.ACTION_REQUEST_ENABLE);
// 请求打开蓝牙设备
        startActivityForResult(enableIntent, REQUEST_ENABLE_BT);
        }
        bluetoothAdapter.startDiscovery();

    }
```

项目四 酒精检测   109

图 4.8   请求打开蓝牙功能

第六步：创建蓝牙进行连接监听，进行蓝牙设备搜索等。具体如代码 CORE0408 所示。

代码 CORE0408   创建蓝牙进行连接所需监听

```
/* 此处添加创建蓝牙进行连接所需监听代码 */
/* 创建一个广播用来监听是否发现设备 */
    private class BluetoothReceiver extends BroadcastReceiver {
            @Override
    public void onReceive(Context context, Intent intent) {
    String action = intent.getAction();
    // 判断蓝牙是否与 action 相等
    if (BluetoothDevice.ACTION_FOUND.equals(action)) {
    device = intent.getParcelableExtra(BluetoothDevice.EXTRA_DEVICE);
                if (isLock(device)) {
                        bluetoothAdapter.cancelDiscovery();
                        if (Matches(device)) {
                            new Thread() {              // 定义线程
                                @Override
                                public void run() {
                                    Message msg = new Message();
                                    msg.what = UPDATE_UI;
                                    myHandler.sendMessage(msg); }
                            }.start();
                } } } } }
```

第七步：对蓝牙发送连接请求，实现蓝牙搜索功能。具体如代码 CORE0409 所示。

代码 CORE0409   对蓝牙发送连接请求

```
/* 此处添加蓝牙发送连接请求代码 */
private class ConnectThread extends Thread {
        private final BluetoothSocket mmSocket;
        private final BluetoothDevice mmDevice;              // 远程设备
        public ConnectThread(BluetoothDevice device) {
```

```java
                mmDevice = device;
                BluetoothSocket tmp = null;
                try {
                        // 获得一个 BluetoothSocket 对象
                        tmp = device.createRfcommSocketToServiceRecord(SPP_
UUID);
                } catch (IOException e) {
                } mmSocket = tmp;
        }
        public void run() {
                setName("ConnectThread");
                mAdapter.cancelDiscovery();          // 取消搜索
                try {
                        mmSocket.connect();
                } catch (IOException e) {
                        connectionFailed();
                        try {
                                mmSocket.close();
                        } catch (IOException e2) {
                        }
                        BluetoothCommService.this.start();
                        return;
                }
                // Reset the ConnectThread because we're done
                synchronized (BluetoothCommService.this) {
                        mConnectThread = null;
                }
                // Start the connected thread，已连接上，管理连接
                connected(mmSocket, mmDevice);
        }
        public void cancel() {
                try {
                mmSocket.close();
                } catch (IOException e) {
                } } }
```

第八步：实现蓝牙匹配功能。具体如代码 CORE0410 所示。

项目四 酒精检测 111

代码 CORE0410 蓝牙匹配

```java
/* 此处添加蓝牙匹配代码 */
/* 是否锁定目标蓝牙 */
    private boolean isLock(BluetoothDevice device) {
        boolean isLockName = (device.getName()).equals(NAME);
        return isLockName;
    }
    /* 蓝牙的配对操作 */
    private boolean Matches(BluetoothDevice device) {
        try {
        if ( device.getBondState() == BluetoothDevice.BOND_NONE) {
        Method creMethod = BluetoothDevice.class.getMethod("createBond");
            creMethod.invoke(device);
            bluetoothFlag = true;
        }
        if (mCommService == null) { // 否则直接尝试连接到蓝牙模块
mCommService = new BluetoothCommService(getActivity(),myHandler);
// 去连接设备用 myHandler 来接受返回的结果
    mCommService.connect(device); // 客户端只负责去连接到指定的设备
        }
        } catch (Exception e) {
        e.printStackTrace();
        }return bluetoothFlag;
    }
```

第九步：实现配对后，进行蓝牙连接。具体如代码 CORE0411 所示。

代码 CORE0411 对蓝牙进行连接

```java
/* 此处添加蓝牙连接代码 * 创建 handle 用来和 UI 做交互 */
    private final Handler myHandler = new Handler() {
        public void handleMessage(Message msg) {
            switch (msg.what) {
            case UPDATE_UI:
                break;
            case MESSAGE_STATE_CHANGE:
                switch (msg.arg1) {
                case BluetoothCommService.STATE_CONNECTED:
Toast.makeText(getActivity(), " 连接成功 ", Toast.LENGTH_LONG).show();
```

```
                    break;
                case BluetoothCommService.STATE_CONNECTING:
        Toast.makeText(getActivity(), " 连接中 ...", Toast.LENGTH_LONG).show();
                    break;
                case BluetoothCommService.STATE_LISTEN:
                    break;
                case BluetoothCommService.STATE_NONE:
        Toast.makeText(getActivity(), " 未连接,请等待连接…", -Toast.LENGTH_
LONG).show();
                    mCommService.start();
                    break;
                    }
                    break;
            case MESSAGE_WRITE:
                    break;
            case MESSAGE_READ:
                    byte[] readBuf = (byte[]) msg.obj;
                    String readMessage = new String(readBuf, 0, msg.arg1);
            String str = String.format(readMessage, new Object[] { "\\r\\n" });
                    inToListView(str);
                    break;
            case MESSAGE_DEVICE_NAME:
                    break;
            case MESSAGE_TOAST:
                    break;
            default:
                    break;
            } } };
```

第十步:连接成功后,进行酒精检测。具体如代码 CORE0412 所示。

代码 CORE0412 进行酒精检测

```
/* 此处添加酒精检测代码 */
private void WriteToProgress(String str) {
        if (str.equals("error")) {
    Toast.makeText(getActivity(), "" + " 请重新点击! ", -Toast.LENGTH_SHORT).
show();
        } else if (str.equals("right")) {
```

```java
            Toast.makeText(getActivity(), "" + " 请对着仪器用力呼气,谢谢合作! ", -Toast.
LENGTH_SHORT).show();
                } else if (str.equals("complete")) {
            Toast.makeText(getActivity(), "" + " 预热完毕! ", Toast.LENGTH_SHORT).show();
                } else if (str.equals("heating")) {
                Toast.makeText(getActivity(), "" + " 系统正在预热中,请稍候! ", Toast.
LENGTH_SHORT).show();
                } else {
                    str = str.trim();
                    char[] b = str.toCharArray();
                    String result = "";
                    for (int i = 0; i < b.length; i++) {
                        if (("0123456789.").indexOf(b[i] + "") != -1) {
                            result += b[i];
                        } }
                    if (!"".equals(result) && result != null) {
                        Double d = Double.valueOf(result.trim());
                        int i = Double.valueOf(d*2200).intValue();
                        if (i < 200) {
            tv_tip.setText(" 您的酒精浓度低于 80% 您可以驾驶,但请注意安全! ");
                        } else {
            tv_tip.setText(" 您的酒精浓度超标,请勿驾车! ");
                        }
                        ProgressBar.setProgress(i);
                        ProgressBar.setTextIsDisplayable(true);
                    } } }
        @Override
        public void onClick(View v) {              // 点击事件
            switch (v.getId()) {
            case R.id.ProgressBar:
                Start();
                break;
            default:
                break;
            } }
        public void Start() {              // 开始线程
            if (mCommService != null) {
            Toast.makeText(getActivity(), " 命令已发送请等待…", Toast.LENGTH_SHORT).show();
```

```
        String max = "start measure\r\n";
        byte[] arrayOfByte1 = CHexConver.hexStr2Bytes(CHexConver.str2HexStr(max));
    mCommService.write(arrayOfByte1);
        } };
```

第十一步：在进度条 RoundProgressBar.java 中设计显示酒精浓度代码。具体如代码 CORE0413 所示，效果如图 4.9 所示。

代码 CORE0413　显示酒精浓度

```
/* 10 此处添加显示酒精浓度代码、画进度百分比 */
        paint.setStrokeWidth(0);
        paint.setColor(textColor);
        paint.setTextSize(textSize);
        paint.setTypeface(Typeface.DEFAULT_BOLD);    // 设置字体
// 中间的进度百分比，先转换成 float 在进行除法运算，不然都为 0
int percent = (int)(((float)progress / (float)max) * 100);
// 测量字体宽度，我们需要根据字体的宽度设置在圆环中间
float textWidth = paint.measureText(percent + "%");
        float str = paint.measureText(" 立即检测 ");
    if(textIsDisplayable && percent != 0 && style == STROKE){
    canvas.drawText(percent + "%", centre - textWidth / 2, centre + textSize/2, paint);
// 画出进度百分比
        }
    if(textIsDisplayable== false){
    canvas.drawText(" 立即检测 ", centre - str / 2, centre + textSize/2, paint);
        }
        /* 画圆弧，画圆环的进度 */
        // 设置进度是实心还是空心
        paint.setStrokeWidth(roundWidth);            // 设置圆环的宽度
        paint.setColor(roundProgressColor);          // 设置进度的颜色
        // 用于定义的圆弧的形状和大小的界限
        RectF oval = new RectF(centre - radius, centre - radius, centre+ radius, centre +radius);
        switch (style) {
        case STROKE:{
            paint.setStyle(Paint.Style.STROKE);
    canvas.drawArc(oval, 0, 360 * progress / max, false, paint); // 根据进度画圆弧
            break;
        }
```

```
case FILL:{
    paint.setStyle(Paint.Style.FILL_AND_STROKE);
    if(progress !=0){
canvas.drawArc(oval, 0, 360 * progress / max, true, paint); // 根据进度画圆弧
        break;
    } }
```

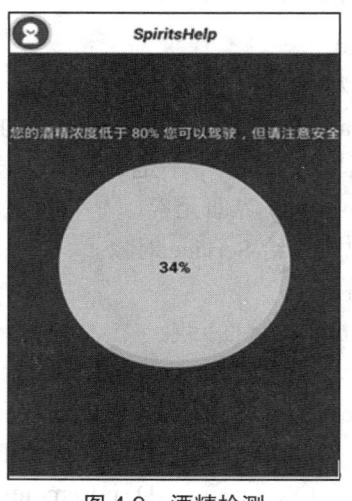

图 4.9　酒精检测

第十二步：运行项目，通过酒精检测仪传回的数据，观察进度条所显示的百分比，判断酒精浓度。如图 4.2 所示。

本项目介绍了 U 酒保酒精检测模块的实现，通过本项目的学习可以了解蓝牙连接的机制和复杂 JSON 解析的方法，掌握圆形进度条的使用方法。实现蓝牙连接检测设备，在界面显示酒精浓度的功能。

Bluetooth	蓝牙	object	对象
notation	记号	Adapter	适配器
sendMessage	发送消息	progerssBar	进度条
connected	已经连接	porizontal	横向的
paint	绘图	thread	线程

一、选择题

1. 下列不属于 Service 生命周期的方法是（　　）。

A.onCreate()　　　　　　B.onDestroy()　　　　　　C.onStop()　　　　　　D. onStart()

2. 下面关于 JSON 说法错误的是（　　）。

A.JSON 是一种数据交互格式

B.JSON 的数据格式有两种为 { } 和 [ ]

C.JSON 数据用 { } 表示 java 中的对象，[ ] 表示 Java 中的 List 对象

D.{“1”:“123”,“2”:“234”,“3”:“345”} 不是 JSON 数据

3. Service 中如何实现更改 Activity 界面元素（　　）。

A. 通过把当前 Activity 对象传递给 Service 对象

B. 通过向 Activity 发送广播

C. 通过 Context 对象更改 Activity 界面元素

D. 可以在 Service 中，调用 Activity 的方法实现更改界面元素

4. 通过调用 BluetoothAdapter 的（　　）方法，可以获取已经配对的蓝牙设备集合。

A.getBoundedDevices()　　　　　　　　B.getRemoteDevice()

C.setBoundedDevices()　　　　　　　　D.setRemoteDevice()

5. 关于 JSON 和 XML 说法，错误的是（　　）。

A. JSON 的速度要远远快于 XML

B. JSON 对数据的描述性比 XML 好

C. JSON 相对于 XML 来讲，数据的体积小

D. JSON 和 XML 同样拥有丰富的解析手段

二、填空题

1.Android 的 JSON 解析部分都在包 org.json 下，主要有以下 5 个类 _____、_____、_____、_____、_____。

2. 本地蓝牙设备的适配类，所有的蓝牙操作都要通过 _____ 类完成。

3. Android 系统提供了两大类进度条样式分别为 _____、_____。

4. Android 平台自带了 JSON 解析的相关 API，可以将文件、输入流中的数据 _____，然后从对象中 _____JSON 保存的数据内容。

5. 圆形进度条需要自定义一个能满足需求的 View，自定义 View 需要继承 View，添加类的构造方法后通过 _____ 方法和自定义属性实现圆形进度条。

三、上机题

1. 设计一个点击按钮开启蓝牙的小程序。

2. 设计一个圆形进度条。

项目五 健康助手

通过 U 酒保项目健康助手模块的实现，了解图片下沉动画的使用方法，掌握如何使用 AsyncTask 类实现更新 UI，学习折线图的构建方法，具有使用异步操作类实现更新 UI 的能力。在任务实现过程中：

● 了解图片下沉动画的使用方法。
● 掌握使用 AsyncTask 类实现更新 UI。
● 掌握折线图的构建方法。

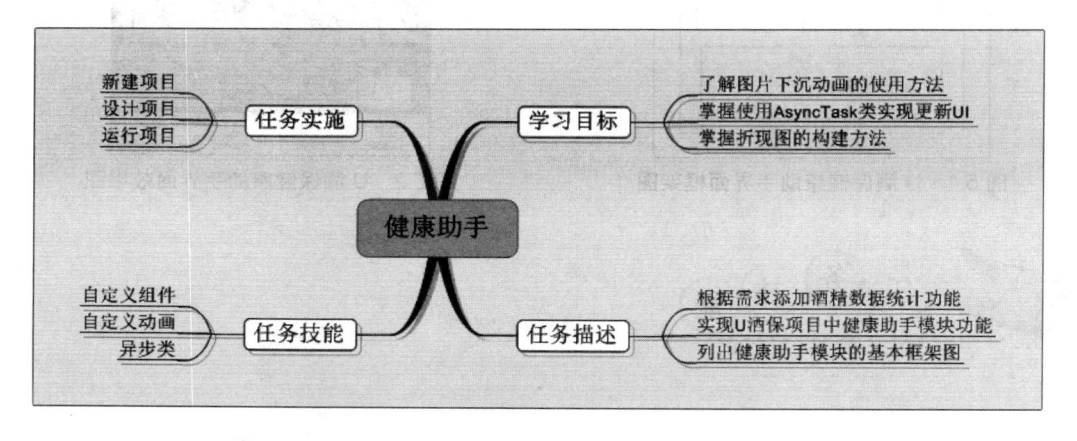

【情境导入】

U 酒保研发团队根据用户的需求设计并开发了酒精数据统计功能，可将历史酒精浓度值以折线图的方式展示，方便用户再次使用时进行查询比对。该软件还设有娱乐模块、微博模块、笑话模块用来丰富业余生活。本项目通过数据监测模块的实现，讲解了折线图以及自定义控件的使用方法。

【功能描述】

本项目将实现 U 酒保项目中健康助手模块功能：
- 实现自定义折线图效果。
- 实现图片点击下沉动画。
- 实现异步加载网络内容。
- 通过跳转传参实现不同页面显示不同数据。

【基本框架】

基本框架如图 5.1 所示。通过本模块的学习，能将框架图 5.1 转换成效果图 5.2 所示。

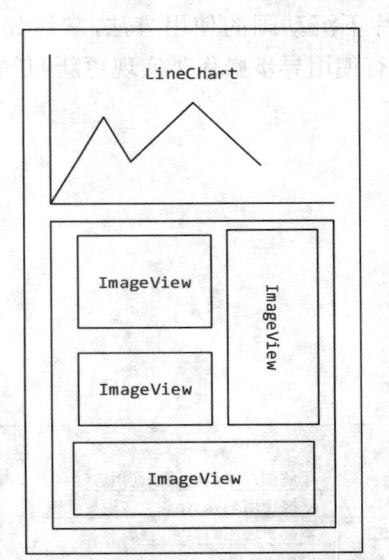

图 5.1　U 酒保健康助手界面框架图

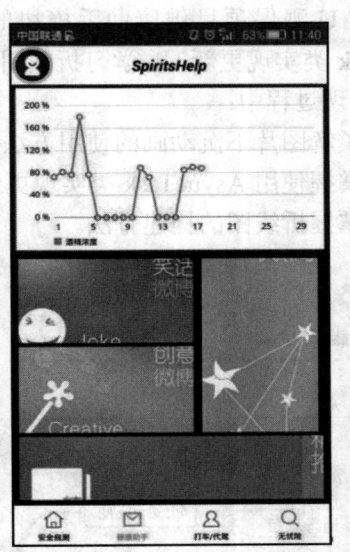

图 5.2　U 酒保健康助手界面效果图

技能点 1　自定义组件

1　自定义组件简介

Android 中所有的 UI 类都以 View 和 ViewGroup 为基础。其中 View 是与用户交互控件的父类，在项目开发过程中，具有重要作用，主要用于在界面中获取矩形区域，完整的实现自定义组件。ViewGroup 是存放 View 的容器，主要用于管理其包含的 View 控件。View 类中提供了多种方法用来构建自定义组件。主要方法如表 5.1 所示。

表 5.1　View 中可被调用的方法

方　　法	说　　明
onFinishInflate()	当应用从 XML 布局文件加载该组件并利用它来构建界面，回调该方法
onMeasure(int width, int heigth)	检测 View 组件及其所包含的所有子组件大小
onLayout(boolean changed, int l, int t, int r, int b)	该组件需要分配子组件的位置、大小时，回调该方法 changed：表示 view 有新的尺寸或位置 l：表示相对于父 view 的 left 位置 t：表示相对于父 view 的 top 位置 r：表示相对于父 view 的 right 位置 b：表示相对于父 view 的 bottom 位置
onSizeChanged(int width, int height, int oldwidth, int oldheigth)	该组件的大小被改变时回调该方法 width：宽度 height：高度 oldwidth：原来宽度 oldheigth：原来高度
onDraw(Canvas canvas)	该组件将要绘制它的内容时回调该方法进行绘制
onKeyDown(int keyCode, keyEvent event)	当某个键被按下时触发该方法
onKeyUp(int keyCode, KeyEvent event)	当松开某个键时触发该方法
onTouchEvent(MotionEvent)	当发生触摸屏事件时触发该方法
onFocusChanged(boolean gainFocus, int direction, Rect rect)	当该组件焦点发生改变时触发该方法 gainFocus：表示该事件的 view 是否获得焦点 direction：表示焦点移动的方向 rect：表示焦点的矩形区域

2　自定义组件方法

View 定义了完整的绘图基本操作，用于自定义组件的构建，在构建过程中，需重写 onMeasure()、onLayout()、onDraw() 方法，具体用法如下所示。

（1）onMeasure() 的用法

绘制 View 前使用 onMeasure() 调用 setMeasuredDimension() 方法对其进行测量，测量流程如图 5.3 所示。

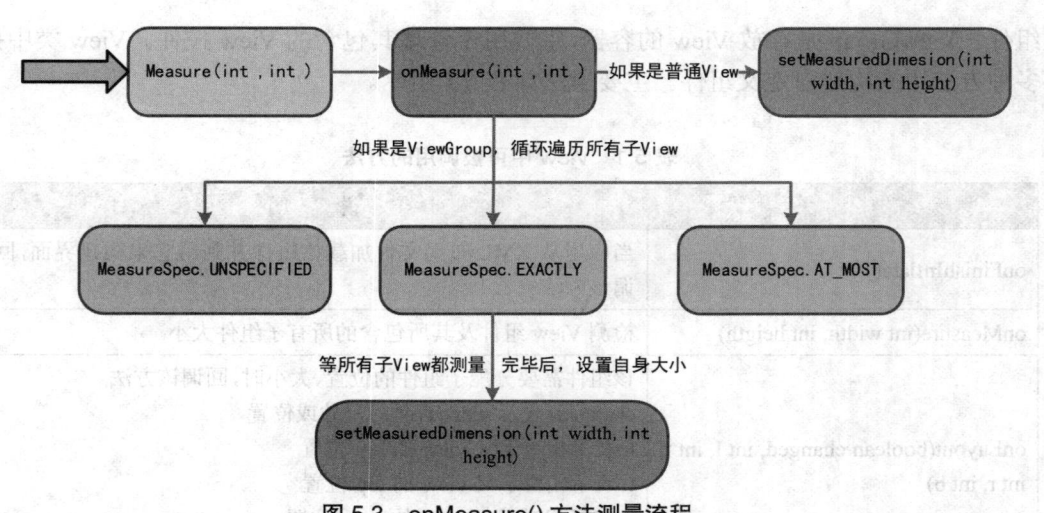

图 5.3　onMeasure() 方法测量流程

由图可知，调用 setMeasuredDimension() 方法有两种途径，分别是：

● View 直接调用 setMeasuredDimension() 方法。

● ViewGroup 调用 MeasureSpec 后调用 setMeasuredDimension() 方法。

其中 MeasureSpec 有三种模式，用于封装父布局传递给子布局的布局要求，如表 5.2 所示。

表 5.2　MeasureSpec 模式

模　式	含　义
UNSPECIFIED	父容器（ViewGroup）对子容器（View）大小没有任何限定
EXACTLY	子容器必须服从父容器为其设定的边界
AT_MOST	父容器根据子容器最大值设置尺寸

当 MeasureSpec 为 EXACTLY 时，重写 onMeasure() 方法，定义 View 的宽度和高度，同时设置文本字体大小，具体代码如下所示。

```
@Override
protected void onMeasure(int widthMeasureSpec, int heightMeasureSpec) {
int widthMode = MeasureSpec.getMode(widthMeasureSpec);   // 定义宽度模式
int widthSize = MeasureSpec.getSize(widthMeasureSpec);    // 定义宽度大小
int heightMode = MeasureSpec.getMode(heightMeasureSpec);  // 定义高度模式
int heightSize = MeasureSpec.getSize(heightMeasureSpec);   // 定义高度大小
int width;
int height ;
   if (widthMode == MeasureSpec.EXACTLY)    {   // 判断宽的模式
      width = widthSize;
} else {
   mPaint.setTextSize(mTitleTextSize);              // 设置文本字体大小
```

项目五 健康助手 121

```
mPaint.getTextBounds(mTitle, 0, mTitle.length(), mBounds);  // 测量 TextView 的位置
    float textWidth = mBounds.width();                    // 获取 TextView 的宽
        int desired = (int) (getPaddingLeft() + textWidth + getPaddingRight());
// 文本高度为距上边距距离 + 文本高度 + 距下边框距离
        width = desired;
} if (heightMode == MeasureSpec.EXACTLY) {              // 判断高的模式
height = heightSize;
    } else {
        mPaint.setTextSize(mTitleTextSize);              // 设置文本字体大小
        mPaint.getTextBounds(mTitle, 0, mTitle.length(), mBounds);  // 获取文本高度
        float textHeight = mBounds.height();
        int desired = (int) (getPaddingTop() + textHeight + getPaddingBottom());
        height = desired;
    } setMeasuredDimension(width, height);
}
```

（2）onLayout() 的用法

onLayout() 主要根据 View 在 Measure() 中测量的大小决定其摆放位置。布局流程如图 5.4 所示。

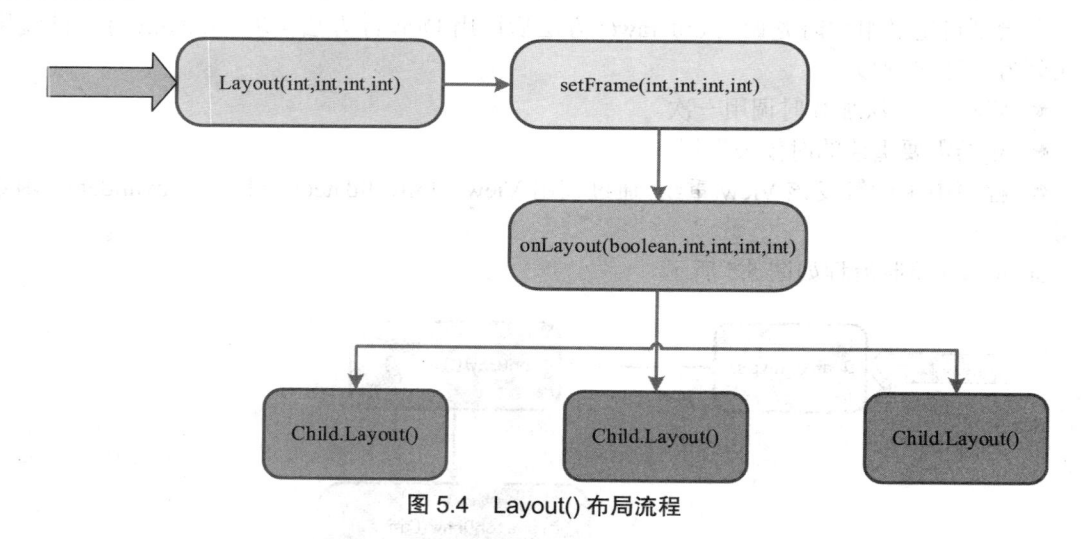

图 5.4　Layout() 布局流程

View 的位置受其他属性影响，如 ViewGroup orientation（方向），gravity（自身），自身的 margin（边缘）等，实现了不同布局的差异。

重写 onLayout() 方法，获得相对于父 View 的位置，遍历所有子视图，并获取 View 大小。具体代码如下所示。

```
@Override
   protected void onLayout(boolean changed, int l, int t, int r, int b) {
      int mTotalHeight = 0;                // 记录总高度
      int childCount = getChildCount();        // 遍历所有子视图
      for (int i = 0; i < childCount; i++) {
      View childView = getChildAt(i);
      // 获取在 onMeasure 中计算的视图尺寸
      int measureHeight = childView.getMeasuredHeight();
      int measuredWidth = childView.getMeasuredWidth();
         childView.layout(l, mTotalHeight, measuredWidth, mTotalHeight + measure-
Height);
      mTotalHeight += measureHeight;
      } }
```

（3）onDraw() 的用法

Canvas 为系统提供一块内存区域,所有绘制都在该内存中进行,绘制完成后系统将布局显示到屏幕中。该 Canvas 对象提供各种绘制点、线、矩形、圆、位图的方法,基本可满足各种绘制要求。

onDraw() 主要是用来把所有 View 都绘制在同一个画布上,达到将 Canvas 内容显示屏幕上。在绘制自定义组件时先调用 onDraw() 方法后调用 Draw() 方法实现。onDraw() 方法调用方式分为三种,分别为:

● View 第一次加载时调用一次。

● 布局需要重绘的时候被调用。

● 程序中手动触发该 View 重绘,通过调用 View 的 invalidate() 函数 postInvalidate() 函数即可。

onDraw() 绘制流程如图 5.5 所示。

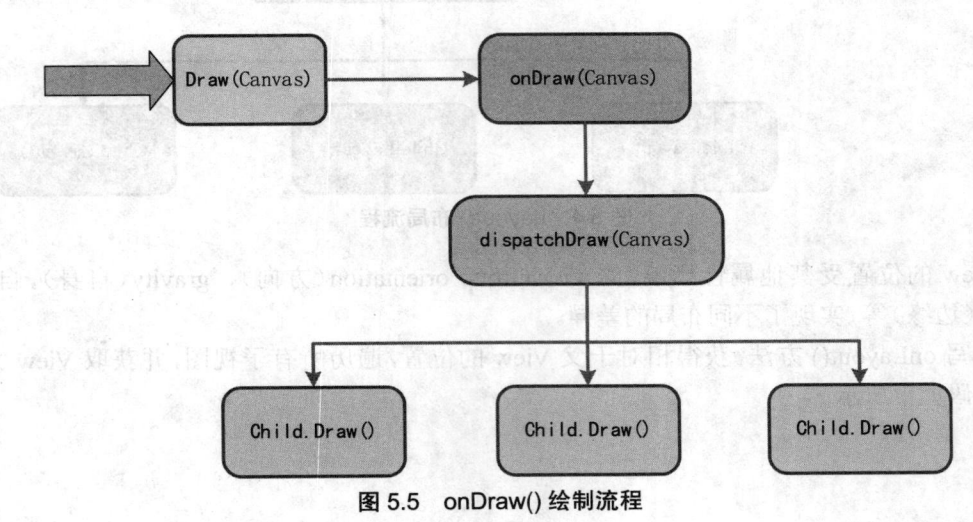

图 5.5　onDraw() 绘制流程

项目五　健康助手

使用 onDraw() 方法设置画笔，并画出 Text 具体在画布中的位置，代码如下所示。

```
@Override
// 画 View
    protected void onDraw(Canvas canvas) {
        mPaint.setColor(Color.YELLOW);        // 设置画笔颜色
        canvas.drawRect(0, 0, getMeasuredWidth(), getMeasuredHeight(), mPaint);
        mPaint.setColor(mTitleTextColor);        // 设置画布
        canvas.drawText(mTitleText, getWidth() / 2 - mBound.width() / 2, getHeight() / 2
+ mBound.height() / 2, mPaint);                // 设置 Text 具体在画布中的位置
    }
```

技能点 2　自定义动画

1　Animation 动画简介

Animation 提供了一系列的动画效果的 API，具有渐变透明度动画效果（alph）、渐变尺寸伸缩动画效果（scale）、画面转换位置移动动画效果（translate）、画面转移旋转动画效果（rotate）等四种动画效果。

2　Animation 方法

Animation 通过对 View 完成一系列图形变换，实现动画效果。定义一组指令，指令指定图形变换类型、触发时间、持续时间。指令可以是以 XML 文件方式定义，也可以是以源代码方式定义。程序沿时间线执行指令便可实现动画效果。

（1）XML 动画

Animation 从总体上可以分为三类：

● Tweened Animation（补间动画），补间动画方式如表 5.3 所示。

表 5.3　补间动画方式

XML 中	含　义
alpha	渐变透明度动画效果
scale	渐变尺寸伸缩效果
translate	画面转换位置动画效果
rotate	画面旋转动画效果

● Frame Animation（逐帧动画），逐帧动画方法如表 5.4 所示。

表 5.4　逐帧动画方法

XML 中	含　义
animation-list	动画列表
item	动画片段

● Property Animation（属性动画），该类 Animation 与补间动画相似，属性动画属性如表 5.5 所示。

表 5.5　属性动画属性

属　性	含　义
android:duration	动画持续时间
android:fillAfter	设置为 true 时，控件动画结束时将保持最后时状态
android:fillBefore	设置为 true 时，控件动画结束时将还原初始状态
android:fillEnabled	与 android:fillBefore 效果相同，都是在动画结束时，将控件还原到初始化状态
android:repeatCount	重复动作次数
android:repeatMode	重复类型，revese：倒序回访，restart：重新播放
android:interpolator	设定插值器

自定义动画在使用过程中需要先定义 XML 资源，将资源放在 /res/anim/ 路径下，如图 5.6 所示，Android Studio 中创建应用时，需要自行创建该路径。

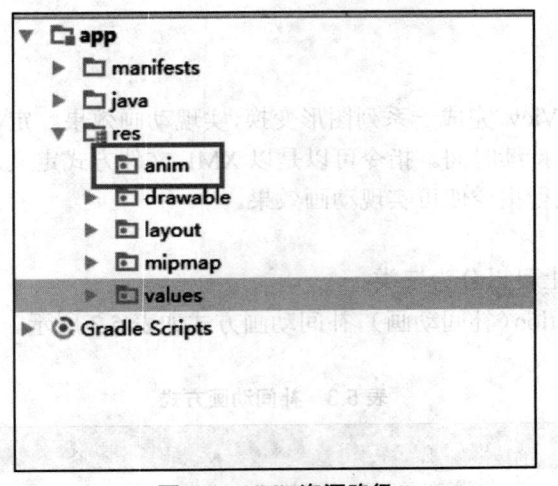

图 5.6　XML 资源路径

定义动画资源文件后，在代码中获取实际的 Animation 对象，可调用 AnimationUtils 的 loadAnimation(Context ctx，int resId) 加载动画资源。

（2）自定义动画

Android 中的图形绘制需要继承 View 组件，并且重写它的 onDraw（Canvas canvas）方法。重写 onDraw(Canvas canvas) 方法时涉及绘图 API：Canvas，Canvas 代表"依附"于指定的 View

画布,提供了如表 5.6 所示的方法。

表 5.6　Canvas 的绘制方法

方　　法	含　　义
drawRect(RectF rect, Paint paint)	绘制区域,参数为 RectF 的区域
drawOval(RectF oval, Paint paint)	绘制矩形的内切椭圆,oval 为矩形区域
drawCircle (float cx, float cy, float radius, Paint paint)	绘制圆形,cx 和 cy 是圆心坐标,radius 是半径长度
void drawArc(RectF oval, float startAngle, float sweepAngle, boolean useCenter, Paint paint)	绘制圆弧形,也是以矩形的内切椭圆为标准。其中,startAngle 为起始角度,sweepAngle 为弧度大小,useCenter 为 true,则是绘制一个扇形,为 false,则只是一段圆弧
drawPath(Path path, Paint paint)	根据给定的 path,绘制连线
drawBitmap(Bitmap bitmap, Rect src, Rect dst, Paint paint)	参数 bitmap 是要进行绘制的 bitmap 对象,参数 src 是指 bitmap 的源区域(一般为 null),dst 是 bitmap 的目标区域,paint 是画笔,可为 null
drawPoint(float x, float y, Paint paint)	根据给定的 x、y 坐标,绘制点
drawLine(float startX, float startY, float stopX, float stopY, Paint paint)	根据给定的起点坐标(startX,startY)和结束点(stopX,stopY)之间绘制连线
drawText(String text, float x, float y, Paint paint)	根据给定的坐标,绘制文字。其中,x 是文本起始的 x 轴坐标,y 是文本纵向结束的 y 轴坐标

　　Canvas 提供的方法提供了另一个 API:Paint,Paint 代表 Canvas 上的画笔,因此 Paint 类主要用于绘制图形,如画笔颜色、画笔笔触粗细、填充颜色等,常用方法如表 5.7 所示。

表 5.7　Paint 的常用方法

方法	含　　义
setARGB(int a, int r, int g, int b)	设置 ARGB 颜色(ARGB 是一种色彩模式,也就是 RGB 色彩模式附加上 Alpha 通道)a:表示透明度 r,g,b:表示 RGB 值
setColor(int color)	设置颜色
setAlpha(int a)	设置透明度
setAntiAlias(boolean aa)	设置是否抗锯齿
setShader(Shader shader)	设置 Paint 的填充效果
setStrokeWidth(float width)	设置 Paint 的笔触宽度
setStyle(Paint.Style style)	设置 Paint 的填充风格
setTextSize(float textSize)	设置绘制文本时的文字大小

　　使用如表 5.7 所示方法制作一个柱状图动画,程序运行后会显示柱状图增长动画,效果如图 5.7 所示。

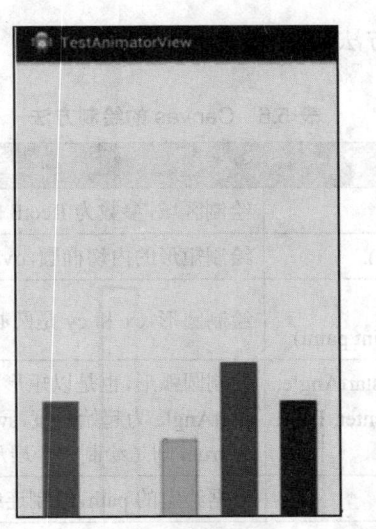

图 5.7 运行效果图

自定义动画实现步骤如下所示：

（1）自定义 View 的派生类 AnimatorView，调用图形数据，并设置画笔及其风格。具体如代码 CORE0501 所示。

代码 CORE0501 创建自定义 View 类继承 View

```java
public class AnimatorView extends View {
private Paint mPaint;
 public AnimatorView(Context context) {                          // 调用图形数据
    super(context);
    nitialize();
}
 public AnimatorView(Context context, AttributeSet attrs) {        // 调用图形数据
    super(context, attrs);
    initialize()
  }
public AnimatorView(Context context, AttributeSet attrs, int defStyle) {
    // 调用图形数据
    super(context, attrs, defStyle);
    initialize();
}
    protected void initialize() {
    mPaint = new Paint();                               // 构造画笔
    mPaint.setAntiAlias(true);
    mPaint.setStyle(Style.FILL);                        // 设置画笔风格类型
  } }
```

项目五　健康助手　　　　　　　　　　　　　　　　　　127

（2）在派生类 AnimatorView 中定义待绘制的图形数据（宽度、间距、速度、颜色、次数）。具体如代码 CORE0502 所示。

代码 CORE0502　定义图形数据

```
public class AnimatorView extends View {
    private static final int RECT_WIDTH = 60;        // 每个矩形块的宽度
    private static final int RECT_DISTANCE = 40;      // 矩形块之间的间距
    private static final int TOTAL_PAINT_TIMES = 100;
// 控制绘制速度，分 100 次完成绘制
    // 待绘制的矩形块矩阵，left 为高度，right 为颜色
    private static final int[][] RECT_ARRAY = {
        {380,Color.GRAY},
        {600,Color.YELLOW},
        {200,Color.GREEN},
        {450,Color.RED},
        {300,Color.BLUE}
    };
    private int mPaintTimes = 0;                      // 当前已经绘制的次数
}
```

（3）通过重载 onDraw() 方法，设置画笔、坐标位置，实现矩形方块的绘制。具体如代码 CORE0503 所示。

代码 CORE0503　重载 onDraw() 函数

```
public class AnimatorView extends View {
    @Override
    protected void onDraw(Canvas canvas) {
        mPaintTimes++;
        for( int i=0; i<RECT_ARRAY.length; i++ ) {      // 画不同颜色的矩形方块
            mPaint.setColor(RECT_ARRAY[i][1]);          // 设置画笔颜色
// 设置 X 轴 Y 轴
            int paintXPos = i*(RECT_WIDTH+RECT_DISTANCE) + RECT_DISTANCE;
            int paintYPos = RECT_ARRAY[i][0]/TOTAL_PAINT_TIMES*mPaintTimes;
            canvas.drawRect(paintXPos,getHeight(), paintXPos+RECT_WIDTH,getHeight()-
paintYPos, mPaint);
        } if( mPaintTimes < TOTAL_PAINT_TIMES ) {
            invalidate();                               // 实现动画的关键点
    } } }
```

128　　　Android 模块化项目实战

（4）在 MainActivity 中遍历 AnimatorView 类，将该动画展示在界面。具体如代码 CORE0504 所示。

代码 CORE0504　在主界面展示自定义 View

```
public class MainActivity extends Activity {
    private AnimatorView mAnimatorView;
    @Override
    protected void onCreate(Bundle savedInstanceState) {
        super.onCreate(savedInstanceState);
        mAnimatorView = new AnimatorView(this);
        setContentView(mAnimatorView);    // 遍历 AnimatorView 类，将动画显示到界面
    } }
```

（5）通过以上步骤实现如图 5.7 自定义动画效果。

　　　　拓展　　大家在学习的过程中有没有考虑过这么一个问题：当我们想制作比较复杂的动画时，例如同时开启多个 view，如果给每个 view 开始动效而不进行优化，很可能会造成 UI 卡顿。那么如何对复杂动画进行优化呢？扫描下方二维码为你呈现优化方式。

技能点 3　异步类

1　AsyncTask 简介

　　AsyncTask 是 Android 提供的轻量级异步类，可直接继承 AsyncTask 在类中实现异步操作，并提供接口反馈当前异步执行程度（可以通过接口实现 UI 进度更新），然后反馈执行结果传递到 UI 主线程。AsyncTask 是封装后的后台任务类，具有结构清晰、功能定义明确，对于多个后台任务时，具有简单，清晰等特点。

2　AsyncTask 类方法

　　AsyncTask 抽象出后台线程运行的五个状态，并提供了五个回调函数：

（1）onPreExecute()：该回调函数在任务被执行后立即由 UI 线程调用。该步骤用来建立任务，在用户接口（UI）上显示进度条（准备运行）。

（2）doInBackground(Params...)：该回调函数由后台线程在 onPreExecute() 方法执行结束后立即调用。在这里执行耗时的后台计算。计算的结果必须由该函数返回，并被传递到 onPostExecute() 中。在该函数内也可以使用 publishProgress(Progress...) 来发布一个或多个进度单位 (unitsof progress)。这些值将会在 onProgressUpdate(Progress...) 中被发布到 UI 线程（后台运行）。

（3）onProgressUpdate(Progress...)：该函数由 UI 线程在 publishProgress(Progress...) 方法调用完后被调用。一般用于动态显示一个进度条（进度更新）。

（4）onPostExecute(Result)：当后台计算结束后调用。后台计算的结果会被作为参数传递给这一函数（完成后台任务）。

（5）onCancelled ()：在调用 AsyncTask 的 cancel() 方法时调用（取消任务）。

AsyncTask 继承自 Object 类，位置为 android.os.AsyncTask。使用 AsyncTask 工作，需要提供三个泛型参数并重载方法（至少重载一个）。AsyncTask 定义的三种泛型类型如表 5.8 所示。

表 5.8　泛型类型

类　　型	含　　义
Params	启动任务执行的输入参数，比如 HTTP 请求的 URL
Progress	后台任务执行的百分比
Result	后台执行任务最终返回的结果，比如 String

使用 AsyncTask 异步加载数据时最少要重写如表 5.9 所示两个方法。

表 5.9　加载数据方法

方法	功能
doInBackground (Params...)	后台执行，比较耗时的操作都可以放在这里。注意这里不能直接操作 UI。此方法在后台线程执行，完成任务的主要工作，通常需要较长的时间。在执行过程中可以调用 publicProgress(Progress...) 来更新任务的进度
onPostExecute (Result)	相当于 Handler 处理 UI 的方式，在这里面可以使用在 doInBackground 得到的结果处理操作 UI。此方法在主线程执行，任务执行的结果作为此方法的参数返回

3　使用 AsyncTask 类示例

使用表 5.9 所示方法，实现倒计时的功能。时长 5 s，当时间结束时在界面显示结束时间并提醒用户倒计时结束，效果如图 5.8 和图 5.9 所示。

图 5.8　倒计时开始效果

图 5.9　倒计时结束效果

实现倒计时功能步骤如下所示：

（1）初始化界面，点击"开始"按钮，调用异步类开始倒计时。具体代码如 CORE0505 所示。

130　Android 模块化项目实战

代码 CORE0505　初始化界面

```
// 初始化界面
bt_button = (Button) findViewById(R.id.btn);
tv_textview = (TextView) findViewById(R.id.text);
// 实现点击功能开始倒计时
bt_button.setOnClickListener(new View.OnClickListener() {
    @Override
    public void onClick(View arg0) {
        // TODO Auto-generated method stub
        new MyAsycnTask(tv_textview).execute(1000);       // 调用异步类
    } });
```

（2）构建 AsyncTask 方法。具体代码如 CORE0506 所示。

代码 CORE0506　初始化界面

```
class MyAsycnTask extends AsyncTask<Integer,Integer,String>{}
```

（3）TestAsyncTask 被后台线程执行后，UI 线程被调用，一般用于初始化界面控件。具体代码如 CORE0507 所示。

代码 CORE0507　开始执行异步线程

```
public MyAsycnTask(TextView tv){
    this.tv = tv;
}
@Override
protected void onPreExecute() {
    // TODO Auto-generated method stub
    super.onPreExecute();
    tv.setText(" 开始执行异步线程 ");
}
```

（4）调用 doInBackground() 方法进行数据的实时获取并将数据进行返回。具体代码如 CORE0508 所示。

代码 CORE0508　实时获取数据

```
@Override
protected String doInBackground(Integer... params) {
    // TODO Auto-generated method stub
    int i = 0;
```

项目五　健康助手

```
for(i=10;i<=50;i+=10){
    try {
        Thread.sleep(1000);
    } catch (InterruptedException e) {
        // TODO Auto-generated catch block
        e.printStackTrace();
    }   publishProgress(i);
}   return i+params[0].intValue()+"";
}
```

（5）将 doInBackground() 方法中返回的数据显示到界面，进行 UI 的实时更新。具体代码如 CORE0509 所示。

代码 CORE0509　将数据显示到界面

```
@Override
protected void onPostExecute(String result) {
    // TODO Auto-generated method stub
    super.onPostExecute(result);
    tv.setText(" 异步线程执行结束 "+result);
}
```

（6）运行项目，实现图 5.8 和图 5.9 所示效果。

拓展　通过上边的学习我们已经掌握 AsyncTask 类的方法并通过练习实现了相应的效果。但是在使用的过程中我们要严格遵守它的准则、明白它的特点、清楚它的局限性。来吧，扫描下方二维码即可了解全部。

通过如下步骤实现如图 5.2 所示 U 酒保健康助手模块。

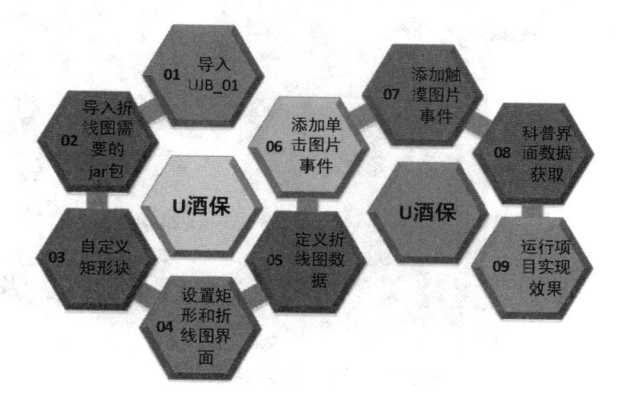

具体步骤如下所示。

第一步：将 UJB_01 导入工程，在其基础上进一步实现 UJB 项目健康助手模块。首先点击 "Open an existing Android Studio project" 打开磁盘路径查找所需项目并导入，如图 5.10 和图 5.11 所示。实现如图 5.12 所示结果图。

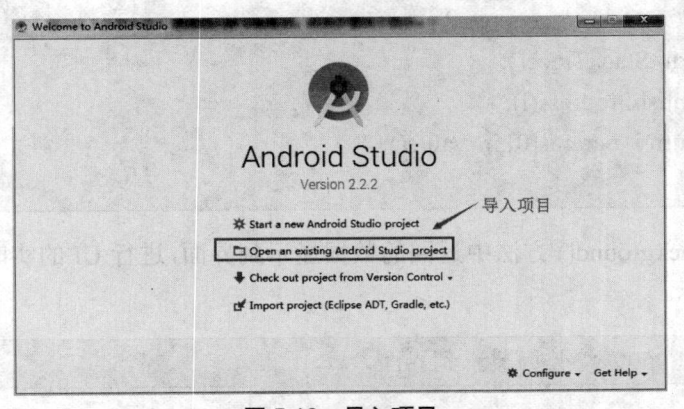

图 5.10　导入项目

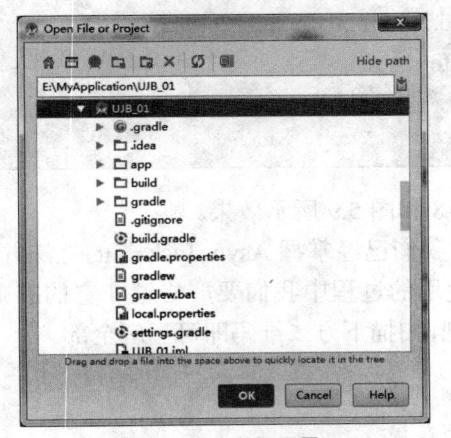

图 5.11　工程目录

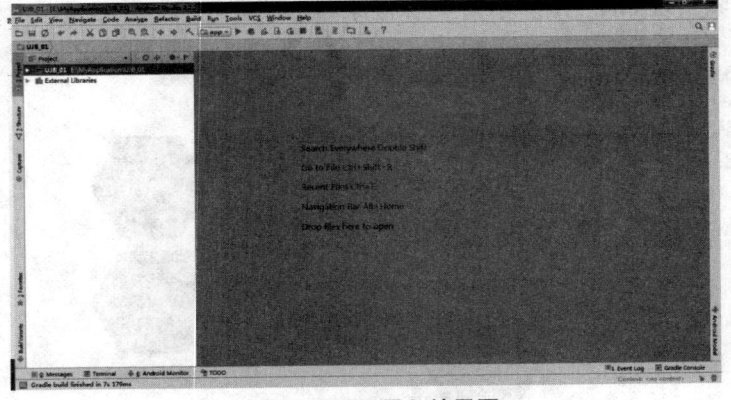

图 5.12　项目导入结果图

第二步：导入实现折线图 MPChartLib.jar，具体步骤如下所示：

（1）复制 MPChartLib.jar 包，添加到如图 5.13 所示目录下。

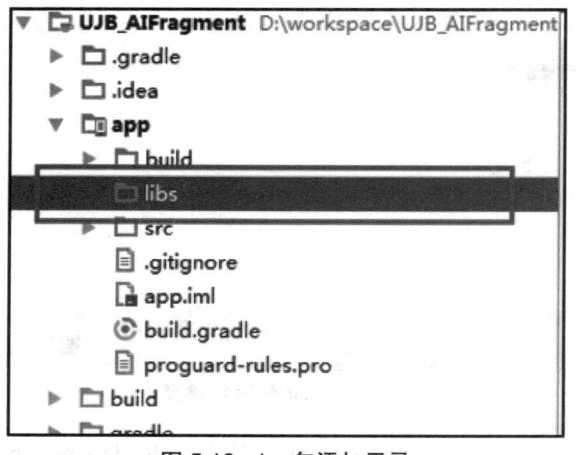

图 5.13　jar 包添加目录

（2）选择该项目，右击鼠标出现如图 5.14 所示界面，点击"Open Module Settings"，跳转到下一界面，选择"app"→"Dependencise"如图 5.15 所示。

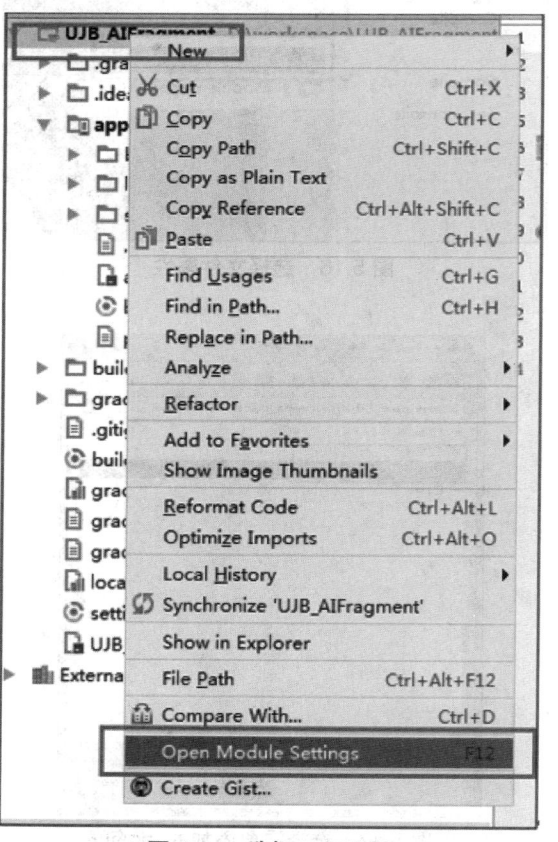

图 5.14　选择工程设置

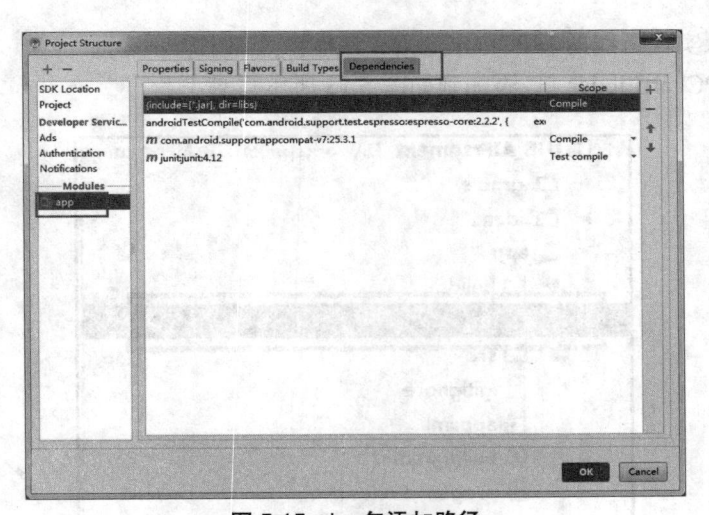

图 5.15　jar 包添加路径

（3）单击"+"显示选择栏,如图 5.16 所示。选择 File dependency,跳转到下一界面,如图 5.17 所示,选择添加的 jar 包后点击"OK"。

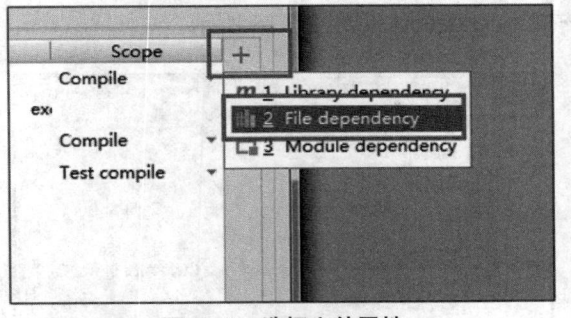

图 5.16　选择文件属性

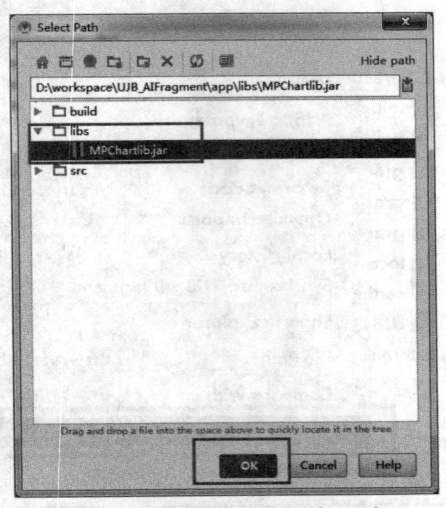

图 5.17　选择需要添加的 jar 包

项目五　健康助手　　135

第三步：使用 onDraw() 方法，新建 MyImageView.java 自定义矩形图片区域，用于放置笑话、娱乐、科普等图片。具体如代码 CORE0510 所示。

```
代码 CORE0510　自定义矩形图片块

/* 此处添加自定义矩形块代码 */
// 画出自定义矩
    @Override
    protected void onDraw(Canvas canvas) {
            super.onDraw(canvas);
            if (isFirst) {
                    isFirst = false;
                    init();
            }
            canvas.setDrawFilter(new  PaintFlagsDrawFilter(0,  Paint.ANTI_ALIAS_
FLAGPaint.FILT
    ER_BITMAP_FLAG));
    }
    public void init() {
        vWidth = getWidth() - getPaddingLeft() - getPaddingRight();      // 得到宽度
        vHeight = getHeight() - getPaddingTop() - getPaddingBottom();   // 得到高度
        Drawable drawable = getDrawable();
        BitmapDrawable bd = (BitmapDrawable) drawable;
        bd.setAntiAlias(true);
    }
```

第四步：实现自定义界面布局，调用 MPChartLib.jar 包与 MyImageView.java 显示矩形图片框和折线图界面 fragment_advise.xml。具体如代码 CORE0511 所示，效果如图 5.18 所示。

```
代码 CORE0511　矩形图片框和折线图界面

/* 此处添加矩形图片框和折线图代码 */
<?xml version="1.0" encoding="utf-8"?>
<LinearLayout xmlns:android="http://schemas.android.com/apk/res/android"
    android:layout_width="match_parent"
    android:layout_height="match_parent"
    android:orientation="vertical" >
<!—定义线性布局展示折线图 -->
<LinearLayout
        android:layout_width="match_parent"
```

```xml
        android:layout_height="wrap_content"
        android:background="#303030"
        android:orientation="vertical"
        android:padding="4dp" >
<!--调用折线图 jar 包绘制折线图 -->
<com.github.mikephil.charting.charts.LineChart
        android:id="@+id/chart"
        android:layout_width="match_parent"
        android:layout_height="200dp"
        android:background="#ffffff"
        android:visibility="visible" />
</LinearLayout>
<!--定义图片框布局 -->
<LinearLayout
        android:layout_width="match_parent"
        android:layout_height="match_parent"
        android:orientation="vertical"
        android:padding="4dp" >
<LinearLayout
        android:layout_width="match_parent"
        android:layout_height="wrap_content"
        android:layout_weight="1"
        android:orientation="horizontal" >
<LinearLayout
            android:layout_width="wrap_content"
            android:layout_height="match_parent"
            android:layout_weight="1"
            android:orientation="vertical" >
<!--导入图片框自定义方法 -->
<com.spirit.lwd.view.MyImageView
            android:id="@+id/tv_one"
            android:layout_width="match_parent"
            android:layout_height="wrap_content"
            android:layout_margin="2dp"
            android:layout_weight="1"
            android:scaleType="matrix"
            android:src="@drawable/ic_click_deflect_left_top" />
<com.spirit.lwd.view.MyImageView
```

```xml
        android:id="@+id/tv_two"
        android:layout_width="match_parent"
        android:layout_height="wrap_content"
        android:layout_margin="2dp"
        android:layout_weight="1"
        android:scaleType="matrix"
        android:src="@drawable/ic_click_deflect_left_bottom" />
</LinearLayout>
<com.spirit.lwd.view.MyImageView
        android:id="@+id/tv_three"
        android:layout_width="wrap_content"
        android:layout_height="match_parent"
        android:layout_margin="2dp"
        android:layout_weight="1"
        android:scaleType="matrix"
        android:src="@drawable/ic_click_deflect_right" />
</LinearLayout>
<com.spirit.lwd.view.MyImageView
        android:id="@+id/tv_four"
        android:layout_width="match_parent"
        android:layout_height="wrap_content"
        android:layout_margin="2dp"
        android:layout_weight="1"
        android:scaleType="matrix"
        android:src="@drawable/ic_click_deflect_bottom" />
</LinearLayout>  </LinearLayout>
```

图 5.18　矩形图片框折线图效果图

138　　　　　　　　　　　　　　Android 模块化项目实战

　　第五步：获取酒精数据 JSON 串并解析，设置折线图属性，并将解析的酒精浓度数据添加到折线图中进行显示。具体如代码 CORE0512 所示，效果如图 5.19 所示。

代码 CORE0512　定义折线图数据

```
/* 为折线图填充数据 */
// 需要填充的数据
String str = "[{ 'density' :73,' date' :0},{ 'density' :82,' date' :1},{ 'density' :77,' date' :2},
{ 'density' :180,' date' :3},{ 'density' :77,' date' :4},{ 'density' :0,' date' :5},{ 'densi-
ty' :0,' date' :6},{ 'density' :0,' date' :7},{ 'density' :0,' date' :8},{ 'density' :0,' date' :9},{ 'densi-
ty' :89,' date' :10},{ 'density' :72,' date' :11},{ 'density' :0,' date' :12},{ 'density' :0,' date' :13},{ 'den-
sity' :0,' date' :14},{ 'density' :85,' date' :15},{ 'density' :90,' date' :16},{ 'density' :88,' date' :17}]";
        Gson gson = new Gson();                    // 解析 JSON
                ArrayList<Data> datas = gson.fromJson(str, new TypeToken<List<Data>>() {
                }.getType());
                appContext.setDatas(datas);
                public View onCreateView(LayoutInflater inflater, ViewGroup container,
Bundle savedInstanceState) {
                mChart.setDrawYValues(false);          // 设置折线图参数
                mChart.setDescription("");
                mChart.setDrawVerticalGrid(false);
                mChart.setDrawGridBackground(false);
                mChart.setUnit(" %");
                mChart.setDrawUnitsInChart(true);
        Typeface mTf = Typeface.createFromAsset(getActivity().getAssets(), "OpenSans-Regu-
lar.ttf");
        XLabels xl = mChart.getXLabels();               // 设置 X 轴属性
                xl.setCenterXLabelText(true);
                xl.setPosition(XLabelPosition.BOTTOM);
                xl.setTypeface(mTf);
                YLabels yl = mChart.getYLabels();          // 设置 Y 轴属性
                yl.setTypeface(mTf);
                yl.setLabelCount(5);
                setData(33, 100);
                mChart.animateX(2500);
                return v;
        }
        private void setData(int count, float range) {
        // 设置折线图参数并将数据进行填充显示到折线图中
```

项目五　健康助手　　　　　　139

```java
ArrayList<String> xVals = new ArrayList<String>();
for (int i = 1; i <= 31; i++) {
    xVals.add((i) + "");
}
ArrayList<Entry> yVals = new ArrayList<Entry>();
for (int i = 0; i < appContext.getDatas().size(); i++) {
    yVals.add(new Entry(appContext.getDatas().get(i).getDensity(), i));
}
LineDataSet set1 = new LineDataSet(yVals, " 酒精浓度 "); // 设置提示信息
set1.setColor(ColorTemplate.getHoloBlue());        // 添加提示信息颜色
set1.setCircleColor(ColorTemplate.getHoloBlue());   // 折线图小圆点颜色
set1.setLineWidth(2f);
set1.setCircleSize(4f);
set1.setFillAlpha(65);
set1.setFillColor(ColorTemplate.getHoloBlue());
set1.setHighLightColor(Color.rgb(255, 0, 0));
ArrayList<LineDataSet> dataSets = new ArrayList<LineDataSet>();
dataSets.add(set1);
LineData data = new LineData(xVals, dataSets);
mChart.setData(data);
}
```

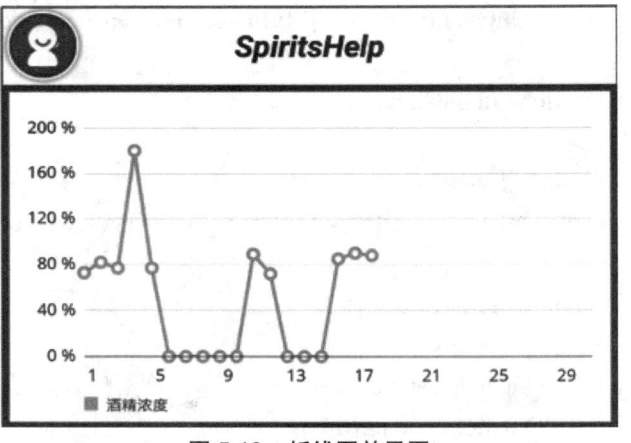

图 5.19　折线图效果图

第六步：使用 OnClick() 方法设置图片点击跳转事件，点击界面中科普、笑话、娱乐等图片，实现对应界面的跳转。具体如代码 CORE0513 所示。

代码 CORE0513 添加单击图片监听

```java
/* 此处添加单击图片监听代码 */
public View onCreateView(LayoutInflater inflater, ViewGroup container, Bundle saved-InstanceState) {
        View v = inflater.inflate(R.layout.fragment_advise, container, false);
        appContext = (AppContext) activity.getApplication();
        mChart = (LineChart) v.findViewById(R.id.chart);          // 获取图片 ID
        tv_one = (MyImageView) v.findViewById(R.id.tv_one);
        tv_two = (MyImageView) v.findViewById(R.id.tv_two);
        tv_three = (MyImageView) v.findViewById(R.id.tv_three);
        tv_four = (MyImageView) v.findViewById(R.id.tv_four);
        // 第一张图片点击效果实现方法
        tv_one.setOnClickIntent(new MyImageView.OnViewClick() {
@Override
            public void onClick() {
            } });
// 第二张图片点击效果实现方法
        tv_two.setOnClickIntent(new MyImageView.OnViewClick() {
@Override
            public void onClick() {
            } });
// 第三张图片点击效果实现方法
        tv_three.setOnClickIntent(new MyImageView.OnViewClick() {
@Override
            public void onClick() {
        } });
// 第四张图片点击效果实现方法
        tv_four.setOnClickIntent(new MyImageView.OnViewClick() {
@Override
            public void onClick() {
            } });
        tv_one.setOnClickListener(this);              // 单击执行方法
        tv_two.setOnClickListener(this);
        tv_three.setOnClickListener(this);
        tv_four.setOnClickListener(this);
        return v;
```

项目五 健康助手 141

```java
        }
    @Override
    public void onClick(View v) {                    // 单击事件进行跳转
            switch (v.getId()) {
            case R.id.tv_one:
                startActivity(new Intent(activity, FirstAdviceActivity.class));
                break;
            case R.id.tv_two:
                startActivity(new Intent(activity, SecondAdviceActivity.class));
                break;
            case R.id.tv_three:
                startActivity(new Intent(activity, ThirdAdviceActivity.class));
                break;
            case R.id.tv_four:
                startActivity(new Intent(activity, FourAdviceActivity.class));
                break;
            default:
                break;
        }    }
```

第七步：使用 onTouchEvent() 方法实现图片触摸事件，当用户触摸图片边缘或中心时，实现图片对应部位的下沉效果。具体如代码 CORE0514 所示，效果如图 5.20 至图 5.22 所示。

代码 CORE0514 添加触摸图片事件

```java
/* 此处添加触摸图片事件代码 */
@Override
    public boolean onTouchEvent(MotionEvent event) {        // 添加触摸事件
            super.onTouchEvent(event);
            if (!onAnimation)
                    return true;
            switch (event.getAction() & MotionEvent.ACTION_MASK) {
            case MotionEvent.ACTION_DOWN:                    // 点击时手指未离开
                    float X = event.getX();
                    float Y = event.getY();
                    RolateX = vWidth / 2 - X;
                    RolateY = vHeight / 2 - Y;
                    XbigY = Math.abs(RolateX) > Math.abs(RolateY) ? true : false;
```

```
                        isScale = X > vWidth / 3 && X < vWidth * 2 / 3 && Y > vHeight /
3&& Y <vHeight * 2 / 3;                              // 下沉程度设置
                isActionMove=false;
                if (isScale) {
                        handler.sendEmptyMessage(1);
                } else {
                        rolateHandler.sendEmptyMessage(1);
                }
                break;
        case MotionEvent.ACTION_MOVE:              // 手指移动
                float x=event.getX();float y=event.getY();
                if(x>vWidth || y>vHeight || x<0 || y<0){
                        isActionMove=true;
                }else{
                        isActionMove=false;
                }
                break;
        case MotionEvent.ACTION_UP:              // 手指移开
                if (isScale) {
                        handler.sendEmptyMessage(6);
                } else {
                        rolateHandler.sendEmptyMessage(6);
                }
                break;
        } eturn true;
}
    public interface OnViewClick {
        public void onClick();
    }
```

图 5.20　图片中心下沉效果图

图 5.21　图片左下沉效果图

项目五　健康助手　　　　143

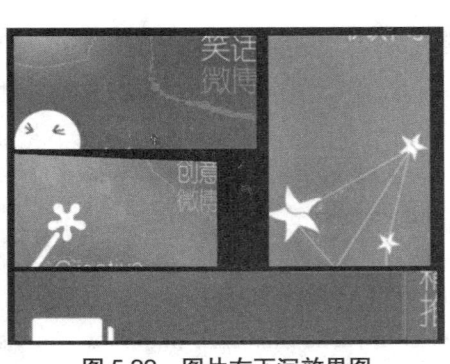

图 5.22　图片右下沉效果图

第八步：点击科普按钮，跳转到科普界面通过异步类获取科普信息，并显示到界面。具体如代码 CORE0515 所示，效果如图 5.23 所示。

代码 CORE0515　科普界面数据获取

```
/* 此处获取科普界面数据 */
class GetUpTask extends AsyncTask<String, Void, Integer> {
        private PullToRefreshListView  mPtrlv;
        public String content = null;
        public GetUpTask() {                              // 重启恢复
        }
        public GetUpTask(PullToRefreshListView  ptrlv) {
            this.mPtrlv = ptrlv;
        }
        @Override
        protected Integer doInBackground(String... params) {
            returnnull;
        }
        @Override
        protectedvoid onPostExecute(Integer result) {       // 异步回调
            super.onPostExecute(result);
            page = page + 1;
            if (NetUtils.isConnected(FirstAdviceActivity.this) == true) {
                getSimulationNews();
            } else {
                getLocalNews();
            }
                mAdapter.notifyDataSetChanged();
                mPtrlv.onRefreshComplete();
}}
```

```java
public void getLocalNews() {              // 获取本地数据并显示到界面
String url = URLCollect.BASE_URL + URLCollect.KNOWLEDGE_URL +"page=
"+page+"&cate=0&pagesize=10";              // 通过路径获取本地数据
        ArrayList<Knowledge> ret = new ArrayList<Knowledge>();
        String data = null;
        try {
            String key = hashKeyForDisk(url);
            DiskLruCache.Snapshot snapShot = mDiskLruCache.get(key);
        if (snapShot != null) {
            data = snapShot.getString(0);
        } else {
            data = "[]";
        }      } catch (IOException e) {
            e.printStackTrace();
        }
        Gson gson = new Gson();              // 进行 JSON 数据的解析
        arrayList = gson.fromJson(data, new TypeToken<List<Knowledge>>() {
        }.getType());
            for (Knowledge news : arrayList) {
            ret.add(news);
        }
        mAdapter.addNews(ret);              // 填充适配器
        mAdapter.notifyDataSetChanged();
    }
public void getSimulationNews() {              // 获取网络数据并显示到界面上
    final String url = URLCollect.BASE_URL + URLCollect.KNOWLEDGE_URL +
"page=" + page + "&cate=0&pagesize=10";              // 通过网络连接获取数据
    // 使用异步进行数据获取
    AsyncHttpHelper.getAbsoluteUrl(url, new TextHttpResponseHandler() {
                @Override
    public void onSuccess(int arg0, Header[] arg1, String data) {
        Gson gson = new Gson();              // 解析 JSON 数据
        ArrayList<Knowledge> mList = gson.fromJson(data, new TypeToken<List
<Knowledge>>() {
                    }.getType());
                    for (Knowledge news : mList) {
                        arrayList.add(news);
                    }
```

项目五　健康助手　　　　　　　　　　　　　　　　　145

```
            mAdapter.addNews(mList);        // 填充适配器
            mAdapter.notifyDataSetChanged();
            WriteToLocal(url, data);
        }
        @Override
public void onFailure(int arg0, Header[] arg1, String data, Throwable arg3) {
        }  });  }
```

酒后喝蜂蜜，能帮助分解酒精，蜂蜜富含果糖、葡萄糖和维生素C，古时候即被用作防止醉酒和消除宿醉的有效食品。尤其是蜂蜜中的葡萄糖有利于直接被人体吸收，对分解酒精有利。

喝芦荟汁，在饮酒之前，如果喝些芦荟汁，对预防酒后头疼和恶心、脸红等症状很有效。此外，芦荟中的枯萎成分芦荟素有健胃作用，可治疗宿醉引起的反胃和恶心等。

图 5.23　科普界面信息效果图

第九步：运行项目，实现如图 5.2 所示效果。

　　本项目主要介绍了 U 酒保健康助手模块的实现。通过自定义组件、自定义动画，实现折线图和图片下沉效果，采用异步请求获取网络数据，并更新 UI。在项目学习过程中可以了解自定义组件和自定义动画的使用方法，掌握异步类的使用原理。

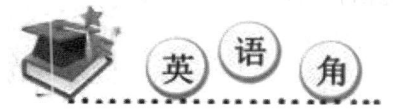

measure	测量	layout	布局
draw	绘制	asyncTask	异步任务
progress	进度	params	参数
group	组	event	事件
detached	独立的	attached	附属的

一、选择题

1. 下列叙述不正确的是（　　　）。

A. 在 Android 开发过程中，View 是开发必不可少的控件

B.View 表示屏幕上的某一块矩形的区域，而且所有的 View 都是矩形的

C. 在 Android 中，控件大致被分为两类，ViewGroup 控件和 View 控件

D. 一个 View 可以有多个父 ViewGroup

2. 对 onDraw(Canvas) 的说明正确的是（　　　）。

A. 该组件需要分配子组件的位置、大小时，回调该方法

B. 该组件将要绘制它的内容时回调该方法进行绘制

C. 该组件的大小被改变时回调该方法

D. 检测 View 组件及其所包含的所有字组件大小

3. 当包含该组件的窗口的可见性发生改变时触发的方法是下列哪一个（　　　）。

A.onWindowFocusChanged(boolean)　　　　　　　B.onAttachedToWindow()

C.onDetachedFromWindow()　　　　　　　　　　　D.onWindowVisibilityChanged(int)

4. Android 系统在绘制 View 之前，必须对 View 进行测量，即告诉系统该画一个多大的 View，这个过程在（　　　）方法中进行。

A.onMeasure()　　　　　　B.onLayout()　　　　　　C.onDraw()　　　　　　D.Measure()

5. 以下说法正确的是（　　　）。

A. 自定义 View 时一般不重写 onDraw() 方法而是重写 Draw() 方法

B. 自定义 View 时一般不重写 Draw() 方法而是重写 onDraw() 方法

C. 自定义 View 时一般重写 Draw() 方法和 onDraw() 方法

D. 自定义 View 时一般不重写 Draw() 方法也不重写 onDraw() 方法

二、填空题

1. ViewGroup 实现了两个接口：_____和_____。

2. 要显示一个 View 那么它所要经历三个方法：_____、_____和绘制_____Draw()。

3. 在调用 onMeasure(int widthSpec, int heightSpec) 方法时，要用到 MeasureSpec，Measure-Spec 有 3 种模式分别是_____、_____和_____。

4. Draw() 方法绘制一定要遵循的顺序：第一步画背景，_____，_____，_____，最后一步画滚动条。

5. Layout() 方法虽然可以被重写，但是不建议去重写，可以直接调用_____方法去确定自身的位置，而且可以去重写_____方法去确定子 View 的位置。

三、上机题

1. 编写代码使用 AsyncTask 异步加载数据。
2. 编写代码显示一个 View。

项目六　打车代驾

通过 U 酒保项目打车代驾模块的实现,了解如何在应用中实现电话服务,掌握 MD5 加密文件的方法及原理,学习如何使用 Stream 流分析软件,具有使用 Stream 流分析软件的能力。在任务实现过程中:

- 了解电话的基本功能。
- 掌握 TelephonyManager 用法。
- 掌握 MD5 加密的方法。
- 掌握 Stream 流使用方法。

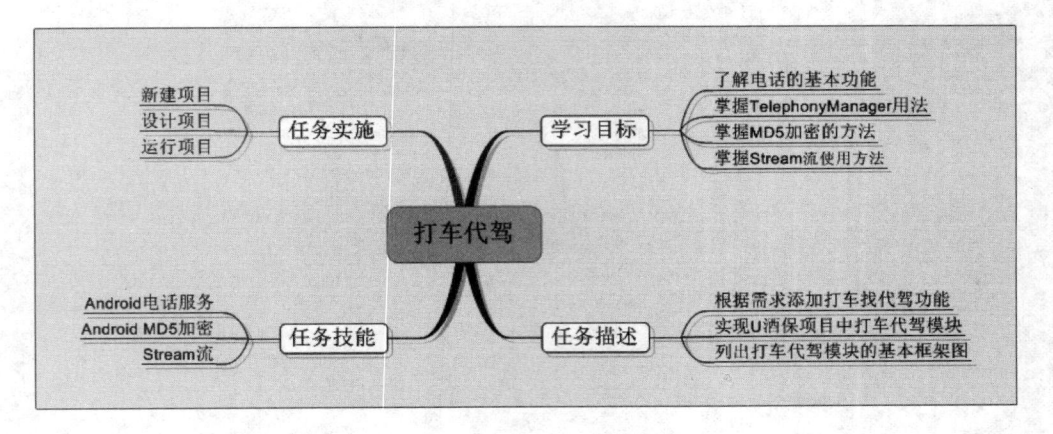

【情境导入】

U 酒保软件根据用户的实际需求进行研发,酒精检测是对酒精浓度的一个可视化显示,当用户检测到自身酒精浓度超标时,可根据情况选择打车或找代驾功能,防止酒驾以及其他安全事故的发生。本项目通过打车代驾模块的实现,讲解了如何实现打车、代驾功能。

【功能描述】

本项目将实现 U 酒保项目中打车代驾模块:

● 实现本地缓存功能。

● 实现界面信息的更新。

● 实现本地文本的加密。

● 实现拨打电话功能。

【基本框架】

基本框架如图 6.1 至图 6.3 所示。通过本模块的学习,能将框架图 6.1 至图 6.3 转换成效果图 6.4 至图 6.6 所示。

图 6.1 主界面框架图   图 6.2 司机列表框架图   图 6.3 司机信息框架图图

图 6.4 主界面效果图   图 6.5 司机列表效果图   图 6.6 司机信息效果图

150　　　　　　　　　　　　　　Android 模块化项目实战

技能点 1　Android 电话服务

1　电话服务简介

电话服务是一款基于 Android 平台的应用。一个电话的基本功能如下：拨叫电话、接听电话、挂断电话、发送短信、网络连接和 PIM 管理。分析 Android 的电话部分，需理解电话实现的背景知识、通讯协议、具体框架。Android 电话服务 API 使应用程序能够访问底层的电话硬件栈，允许创建自己的拨号程序和电话状态监视的功能，并集成到应用程序中。Android 的电话系统构成如图 6.7 所示。

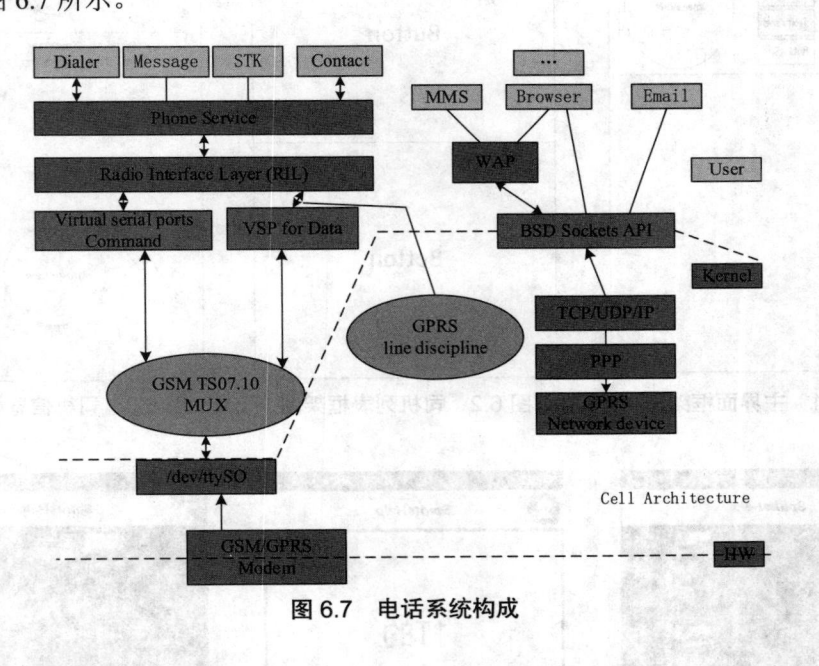

图 6.7　电话系统构成

2　TelephonyManager 用法

Android 提供的系统服务 TelephonyManager（电话管理器），TelephonyManager 用于管理手机通话状态、获取电话信息（设备信息、SIM 卡信息以及网络信息）、侦听电话状态（呼叫状态服务状态、信号强度状态等）以及可以调用电话拨号器拨打电话。TelephonyManager 的相关属性如表 6.1 所示。

TelephonyManager 的相关方法如表 6.2 所示。

项目六　打车代驾

表 6.1　TelephonyManager 的相关属性

类　型	属　　性	说　　明
String	ACTION_PHONE_STATE_CHANGED	指设备上的呼叫状态已改变
int	CALL_STATE_IDLE	无呼入或已挂机
int	CALL_STATE_OFFHOOK	有呼入
int	CALL_STATE_RINGING	接听中
int	DATA_ACTIVITY_DORMANT	电话数据活动状态类型:睡眠模式(3.1 版本)
int	DATA_ACTIVITY_IN	电话数据活动状态类型:数据流入
int	DATA_ACTIVITY_INOUT	电话数据活动状态类型:数据交互
int	DATA_ACTIVITY_NONE	电话数据活动状态类型:无数据流动
int	DATA_ACTIVITY_OUT	电话数据活动状态类型:数据流出
int	DATA_CONNECTED	数据连接状态类型:已连接
int	DATA_CONNECTING	数据连接状态类型:正在连接
int	DATA_DISCONNECTED	数据连接状态类型:断开
int	DATA_SUSPENDED	数据连接状态类型:已暂停

表 6.2　TelephonyManager 的相关方法

类　型	方　法	说　明
int	getCallState()	返回一个数,指示设备上的呼叫状态
CellLocation	getCellLocation()	返回设备的当前位置
int	getDataActivity()	电话数据活动状态类型定义在 TelephoneyManager 类中
int	getDataState()	返回一个常量,指示当前数据连接状态
String	getDeviceId()	获取设备标识(IMEI)
String	getNetworkOperator()	获取 SIM 移动国家代码(MCC)和移动网络代码(MNC)
String	getNetworkOperator-Name()	获取服务提供商姓名(中国移动、中国联通等)
int	getNetworkType()	获取网络类型
String	getSimCountryIso()	获取 SIM 卡中国家 ISO 代码
String	getSimOperator()	获得 SIM 卡中移动国家代码(MCC)和移动网络代码(MNC)
String	getNetworkOperator-Name()	获取服务提供商姓名(中国移动、中国联通等)
int	getNetworkType()	获取网络类型

TelephonyManager 服务的实现:

（1）TelephonyManager 管理电话服务 API 的访问,获取 TelephonyManager 服务对象,具体代码如下所示。

```
TelephonyManagertManager=(TelephonyManager)getSystemService(Context.TELE-
PHONY_SERVICE);
```

（2）添加控制和读取通话的权限，具体代码如下所示。

```
<!-- 授予该应用控制通话的权限 -->
<uses-permission android:name="android.permission.CALL_PHONE">
</uses-permission>
<!-- 授予该应用读取通话状态的权限 -->
<uses-permission android:name="android.permission.READ_PHONE_STATE">
</uses-permission>
```

技能点 2　Android MD5 加密

1　MD5 简介

MD5 的全称 Message-Digest Algorithm 5（信息 - 摘要算法），90 年代初由 MIT Laboratory for Computer Science 和 RSA Data Security Inc 的 Ronald L. Rivest 开发出来，经 MD2、MD3 和 MD4 发展而来，是单向加密算法。其作用是把一个任意长度的字节串变成一定长的大整数。无论是 MD2、MD4 还是 MD5，都需要获得一个随机长度的信息并产生一个 128 位的信息摘要。

MD5 值如文件的"数字指纹"。每个文件的 MD5 值是不同的，如果对文件做了改动，其 MD5 值也就是对应的"数字指纹"就会发生变化。如下载服务器针时对一个文件预先提供一个 MD5 值，用户下载完该文件后，用这个算法会重新计算下载文件的 MD5 值，通过比较这两个值是否相同，就能判断下载的文件是否出错，或文件是否被篡改了。利用 MD5 算法来进行文件校验的方案被大量应用到软件下载站、论坛数据库、系统文件安全等方面。

2　MD5 加密原理及特点

MD5 加密算法分析：MD5 以 512 位分组来处理输入的信息，每一分组被划分为 16 个 32 位子分组，经过了一系列处理后，算法输出由四个 32 位分组组成，将这四个 32 位分组级联后将生成一个 128 位散列值。MD5 加密有以下几个特点。
- 压缩性：任意长度的数据，算出的 MD5 值长度都是固定的。
- 容易计算：从原数据计算出 MD5 值很容易。
- 抗修改性：对原数据进行任何改动，所得到的 MD5 值都有很大区别。
- 强抗碰撞：已知原数据和其 MD5 值，再找到一个具有相同 MD5 值的数据（即伪造数据）是非常困难的。

项目六　打车代驾　　153

3　MD5 加密算法的实现

（1）在加密之前要计算字符串的 MD5 值，具体代码如下所示。

```java
public static String md5(String string) {
    if (TextUtils.isEmpty(string)) {
        return "";
    }
    MessageDigest md5 = null;            // 输入不为空
    try {
    md5 = MessageDigest.getInstance("MD5");            // 得到一个 MessageDigest 对象
    // 通过调用 .digest(byte[]) 得到了加密后的字节数组
    byte[] bytes = md5.digest(string.getBytes());
        String result = "";
    for (byte b : bytes) {                // 枚举查出数据
        String temp = Integer.toHexString(b & 0xff);
        if (temp.length() == 1) {
        temp = "0" + temp;
        } result += temp;
        } return result;
        } catch (NoSuchAlgorithmException e) {
        e.printStackTrace();
    } return "";
}
```

（2）计算文件的 MD5 值，具体代码如下所示。

```java
public static String md5(File file) {
    if (file == null || !file.isFile() || !file.exists()) {
        return "";
    }
    FileInputStream in = null;
    String result = "";
    byte buffer[] = new byte[8192];
    int len;
    try {
// 创建 MD5 转换器和文件流
MessageDigest md5 = MessageDigest.getInstance("MD5");
    in = new FileInputStream(file);
    while ((len = in.read(buffer)) != -1) {
```

```
        md5.update(buffer, 0, len);
      }
      byte[] bytes = md5.digest();
      for (byte b : bytes) {
        String temp = Integer.toHexString(b & 0xff);
        if (temp.length() == 1) {
          temp = "0" + temp;
        } result += temp;
      } } catch (Exception e) {
        e.printStackTrace();
    }finally {
      if(null!=in){
        try {
          in.close();
        } catch (IOException e) {
          e.printStackTrace();
      } } }
    return result;
  }
```

（3）MD5 多次加密。

MD5 加密本身是不可逆的,但可破译,有关 MD5 解密的网站数不胜数,破解机制采用穷举法,就是跑字典。为了加大 MD5 的破解难度,可以采用对字符串进行多次加密处理。具体代码如下所示。

```
    public static String md5(String string, int times) {
        if (TextUtils.isEmpty(string)) {
            return "";
        }
        String md5 = md5(string);
        for (int i = 0; i < times - 1; i++) {
            md5 = md5(md5);
        } return md5(md5);
    }
```

拓展　MD5 是计算机安全领域广泛使用的一种散列函数,用以提供消息的完整性保护。MD5 的前身有 MD2、MD3 和 MD4。想要详细了解它们各自的发展史以及大神对 MD5 的探索事迹吗? 万能的二维码会为你呈现精彩内容。

项目六　打车代驾　155

技能点 3　Stream 流

1　Stream 流简介

Stream 流是指数据传输时的形态，Java 为 Stream 流提供了多个内置类，如 IO 输入流、输出流。流从功能上分为两大类：节点流类、过滤流类（也叫处理流类）。程序直接操作目标设备所对应的类叫节点流类。程序通过间接流类调用节点流类读取不同类型的数据叫过滤流类，也称为包装类。

2　方法说明

在开发中，流是一种常见的形态，如文件的输入输出，都需要以流的形态进行操作。在操作流之前首先要生成流，流生成方法如表 6.3 所示。

表 6.3　流生成方法

方　　法	描　　述
Collection.stream()	使用一个集合的元素创建一个流
Stream.of(T)	使用传递给工厂方法的参数创建一个流
Stream.of(T[])	使用一个数组的元素创建一个流
Stream.empty()	创建一个空流
Stream.iterate (Tfirst,BinaryOperator<T>f)	创建一个包含序列 first, f(first), f(f(first)), ... 的无限流
Stream.iterate(Tfirst,Predi-cate<T>test,BinaryOperator<T>f)	（仅限 Java 9）类似于 Stream.iterate(Tfirst, BinaryOperator<T> f)，但流在测试预期返回 false 的第一个元素上终止
Stream.generate(Supplier<T>f)	使用一个生成器函数创建一个无限流
IntStream.range(lower,upper)	创建一个由下限到上限（不含）之间的元素组成的 IntStream
IntStream.rangeClosed(lower,upper)	创建一个由下限到上限（含）之间的元素组成的 IntStream
BufferedReader.lines()	创建一个有来自 BufferedReader 的行组成的流
BitSet.stream()	创建一个由 BitSet 中设置位的索引组成的 IntStream
Stream.chars()	创建一个与 String 中的字符对应的 IntStream

中间操作负责将一种类型的流转换为另一种类型的流，调用中间操作只会设置流管道的下一个阶段，不会启动任何操作。中间操作可分为无状态和有状态操作。无状态操作（比如 filter() 或 map()）可独立处理每个元素，有状态操作可以对之前影响其他元素处理的元素状态进行合并。

流的中间操作具体如表 6.4 所示。

Android 模块化项目实战

表 6.4　中间流操作

操　作	内　容
filter(Predicate\<T>)	与预期匹配的流的元素
map(Function\<T, U>)	将提供的函数应用于流的元素的结果
flatMap(Function\<T, Stream\<U>>	将提供的流处理函数应用于流元素后获得的流元素
distinct()	已删除了重复的流元素
sorted()	按自然顺序排序的流元素
Sorted(Comparator\<T>)	按提供的比较符排序的流元素
limit(long)	截断至所提供长度的流元素
skip(long)	丢弃了前 N 个元素的流元素
takeWhile(Predicate\<T>)	（仅限 Java 9）在第一个提供的预期不是 true 的元素处阶段的流元素
dropWhile(Predicate\<T>)	（仅限 Java 9）丢弃了所提供的预期为 true 的初始元素分段的流元素

执行终止操作时，会终止流管道，如果想再次编辑同一个数据集，可以设置一个新的流管道。数据集的处理在执行终止操作时开始，比如缩减（min() 或 max()）、应用（forEach()）或搜索（findFirst()）操作。流的终止操作如表 6.5 所示。

表 6.5　终止流操作

操　作	描　述
forEach(Consumer\<T> action)	将提供的操作应用于流的每个元素
toArray()	使用流元素创建一个数组
reduce(...)	将流的元素聚合为一个汇总值
collect(...)	将流的元素聚合到一个汇总结果容器中
min(Comparator\<T>)	通过比较符返回流的最小元素
max(Comparator\<T>)	通过比较符返回流的最大元素
count()	返回流的大小
{any,all,none}Match(Predicate\<T>)	返回流的任何 / 所有元素是否与提供的预期相匹配
findFirst()	返回流的第一个元素（如果有）
findAny()	返回流的任何元素（如果有）

3　Stream 流实现步骤

通过一个九宫格的小游戏，体现 Stream 流的操作机制，运行效果如图 6.8 所示。

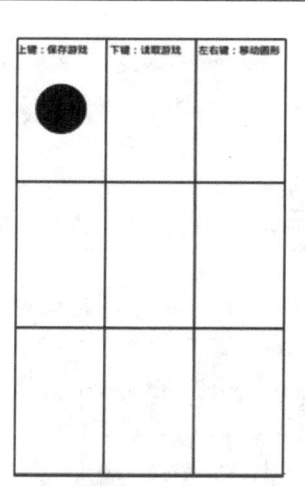

图 6.8　StreamProject 运行效果

实现步骤如下所示：

（1）初始化函数（画笔、颜色等）。具体代码如 CORE0601 所示。

代码 CORE0601　初始化函数

```
public MySurfaceView(Context context) {
        super(context);
        sfh = this.getHolder();
        sfh.addCallback(this);
        paint = new Paint();
        paint.setColor(Color.BLACK);
        paint.setAntiAlias(true);
        setFocusable(true);
}
```

（2）通过 myDraw() 方法将游戏界面九等分，并根据圆形下标位置将其绘制到相应的方格中具体代码如 CORE0602 所示。

代码 CORE0602　对游戏界面的图形进行绘制

```
public void myDraw() {
        try {
                canvas = sfh.lockCanvas();
                if (canvas != null) {
                        canvas.drawColor(Color.WHITE);
                        paint.setColor(Color.BLACK);
                        paint.setStyle(Style.STROKE);
                        // 绘制九宫格（将屏幕九等份）
```

```java
                    int tileW = screenW / 3;              // 得到每个方格的宽高
                    int tileH = screenH / 3;
                for (int i = 0; i < 3; i++) {
                    for (int j = 0; j < 3; j++) {
        canvas.drawRect(i * tileW, j * tileH, (i + 1) * tileW, (j + 1) * tileH, paint);
                        } }
            // 根据得到的圆形下标位置，绘制到相应的方格中
            paint.setStyle(Style.FILL);
            canvas.drawCircle(creentTileIndex % 3 * tileW + tileW / 2, creentTileIndex /
3 * tileH + tileH / 2, 30, paint);
            canvas.drawText(" 上键：保存游戏 ", 0, 20, paint);   // 操作说明
            canvas.drawText(" 下键：读取游戏 ", 110, 20, paint);
            canvas.drawText(" 左右键：移动圆形 ", 215, 20, paint);
                }
            } catch (Exception e) {
                // TODO: handle exception
            } finally {
                if (canvas != null)
                    sfh.unlockCanvasAndPost(canvas);
            } }
```

（3）设置触屏监听，当触屏时调用该方法。具体代码如 CORE0603 所示。

代码 CORE0603　设置触屏监听

```java
@Override
    public boolean onTouchEvent(MotionEvent event) {
            return true;
    }
```

（4）设置按键监听，使用输出和输入流在 SD 卡中存取游戏状态。具体代码如 CORE0604
所示。

代码 CORE0604　设置按键监听

```java
@Override
    public boolean onKeyDown(int keyCode, KeyEvent event) {
        // 用到的读出、写入流
        FileOutputStream fos = null;
        FileInputStream fis = null;
        DataOutputStream dos = null;
```

项目六 打车代驾

```
            DataInputStream dis = null;
    // 上键保存游戏状态
    if (keyCode == KeyEvent.KEYCODE_DPAD_UP) {
        try {
                    /* 从 SDcard 中写入数据
                    试探终端是否有 sdcard! 并且探测 SDCard 是否处于被移除的
状态 */
            if (Environment.getExternalStorageState() != null && !Environment.getExternal
            StorageState().equals("removed")) {
                    File path = new File("/sdcard/himi");          // 创建目录
                    File f = new File("/sdcard/himi/save.himi");        // 创建文件
    // 下键读取游戏状态
    } else if (keyCode == KeyEvent.KEYCODE_DPAD_DOWN) {
                    boolean isHaveSDCard = false;
    /* 从 SDcard 中读取数据
    试探终端是否有 sdcard! 并且探测 SDCard 是否处于被移除的状态 */
    if (Environment.getExternalStorageState() != null && !Environment.getExter-
nal
    StorageState().equals("removed")) {
            isHaveSDCard = true;
                }
```

（5）调用 myDraw() 方法和 logic() 方法实现游戏的逻辑，具体代码如 CORE0605 所示。

代码 CORE0605 游戏实现的逻辑

```
private void logic() {
    }
    @Override
    public void run() {
            while (flag) {
                    long start = System.currentTimeMillis();
                    myDraw();
                    logic();
                    long end = System.currentTimeMillis();
                    try {
                            if (end - start < 50) {
                                    Thread.sleep(50 - (end - start));
                            }         } catch (InterruptedException e) {
```

160 　　　　　　　　　　　　Android 模块化项目实战

```
                        e.printStackTrace();
        }            } }
```

（6）响应函数，实例线程并设置启动线程，具体代码如 CORE0606 所示。

代码 CORE0606　响应函数

```
public void surfaceCreated(SurfaceHolder holder) {
        screenW = this.getWidth();
        screenH = this.getHeight();
        flag = true;
        th = new Thread(this);              // 实例线程
        th.start();                         // 启动线程
    }
```

（7）运行程序，效果如图 6.8 所示。

拓展　作为一个程序员想必大家都知道，我们所写的程序是通过流来完成输入和输出的。尽管程序链接的物理设备不尽相同，所以流的行为方式具有相同的方式。想要对流进行进一步的了解以及清楚流的分类，可以扫描下方二维码。

通过如下步骤实现 U 酒保的打车代驾模块。

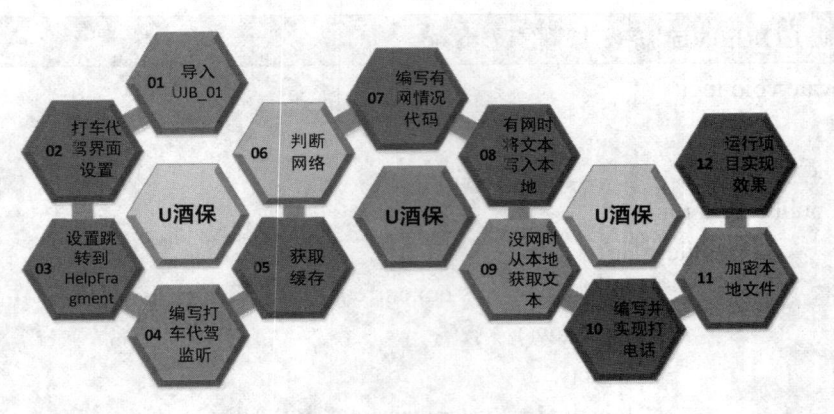

具体步骤如下所示。

第一步：将 UJB_01 导入工程，在其基础上进一步实现 UJB 项目打车代驾模块。首先点击"Open an existing Android Studio project"打开磁盘路径查找所需项目并导入，如图 6.9 和图 6.10 所示。实现如图 6.11 所示结果图。

项目六　打车代驾　　　161

图 6.9　导入项目

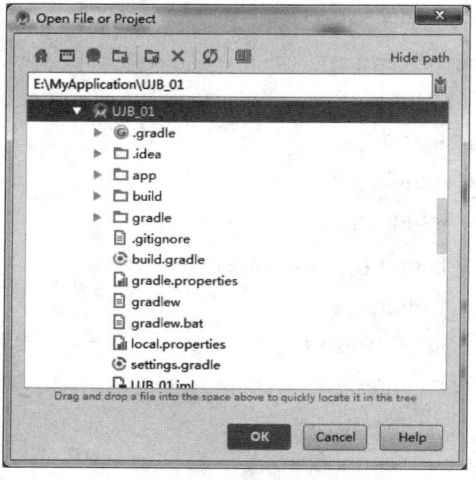

图 6.10　工程目录

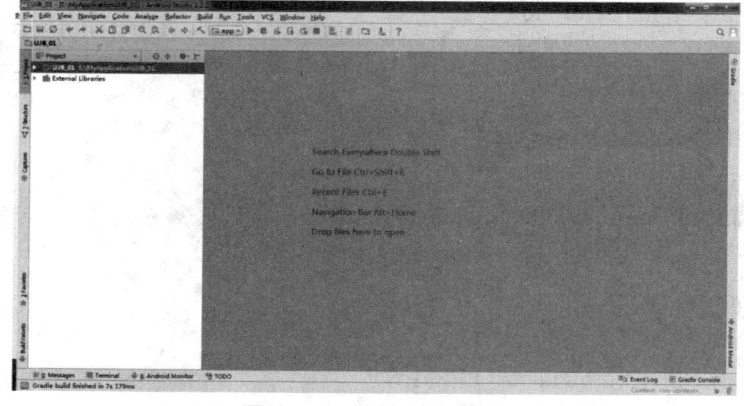

图 6.11　项目导入结果图

第二步：打车代驾界面布局格式。如图 6.4 所示。具体如代码 CORE0607 所示。

代码 CORE0607 编写打车代驾页面布局

```xml
/* 此处添加打车代驾界面布局代码 */
<?xml version="1.0" encoding="utf-8"?>
<LinearLayout xmlns:android="http://schemas.android.com/apk/res/android"
    xmlns:tools="http://schemas.android.com/tools"
    xmlns:ripple="http://schemas.android.com/apk/res/com.spirit.lwd"
    android:layout_width="match_parent"
    android:layout_height="match_parent"
    android:background="#303030"
    android:orientation="vertical" >
<LinearLayout
    android:layout_width="match_parent"
    android:layout_height="match_parent"
    android:orientation="vertical"
    android:padding="4dp" >
<com.spirit.lwd.view.RippleView
    android:id="@+id/iv_dache"
    android:layout_width="match_parent"
    android:layout_height="wrap_content"
    android:layout_weight="1"
    android:background="#FFFFFF"
    android:gravity="center"
    android:text="@string/Taxis"
    android:textColor="#000000"
    android:textSize="40sp"
    ripple:alphaFactor="0.0"
    ripple:hover="true"
    ripple:rippleColor="#778899" />
<com.spirit.lwd.view.RippleView
    android:id="@+id/iv_daijia"
    android:layout_width="match_parent"
    android:layout_height="wrap_content"
    android:layout_marginTop="4dp"
    android:layout_weight="1"
    android:background="#FFFFFF"
    android:gravity="center"
```

项目六　打车代驾　　　　　　　　　　　　　　　　　　　　　　163

```
        android:text="@string/ChauffeurDrive"
        android:textColor="#000000"
        android:textSize="40sp"
        ripple:alphaFactor="0.0"
        ripple:hover="true"
        ripple:rippleColor="#778899" />
    </LinearLayout>  </LinearLayout>
```

第三步：在 src 文件夹下建立 MainActivity.java 文件中，实现点击"打车 / 代驾"小标，跳转到 HelpFragment。具体如代码 CORE0608 所示。

代码 CORE0608　编写跳转到 HelpFragment

```
/* 此处添加点击打车代驾图标实现跳转到打车代驾 HelpFragment 功能代码 */
public class MainActivity extends FragmentActivity implements OnClickListener {
    public static final String TAG_1 = "HOME";
    public static final String TAG_2 = "ADVISE";
    public static final String TAG_3 = "HELP";
    public static final String TAG_4 = "INSURANCE";
@Override
    protected void onCreate(Bundle savedInstanceState) {
        super.onCreate(savedInstanceState);
        setContentView(R.layout.activity_tabs);
        PushAgent mPushAgent = PushAgent.getInstance(this.getApplication());
        mPushAgent.enable();
        mTabHost = (TabHost) findViewById(android.R.id.tabhost);
        mTabHost.setup();
        mTabManager = new TabManager(this, mTabHost, R.id.realtabcontent);
        mTabManager.addTab(mTabHost.newTabSpec(TAG_1).setIndicator(cre-
ateTabView(" 安全自测 ", R.drawable.tabbar_home_selector)), HomeFragment.class, null);
        mTabManager.addTab(mTabHost.newTabSpec(TAG_2).setIndicator(cre-
ateTabView(" 健康助手 ", R.drawable.tabbar_msg_selector)), AdviseFragment.class, null);
        mTabManager.addTab(mTabHost.newTabSpec(TAG_3).setIndicator(cre-
ateTabView(" 打 车 / 代 驾 ",  R.drawable.tabbar_profile_selector)),  HelpFragment.class,
null);
        mTabManager.addTab(mTabHost.newTabSpec(TAG_4).setIndicator(cre-
ateTabView(" 无忧险 ", R.drawable.tabbar_discover_selector)), AIFragment.class, null);
```

第四步：对打车和代驾按钮设置监听，并判断是否连接网络，实现有网的情况下点击后可进入打车代驾界面的功能。具体如代码 CORE0609 所示。

代码 CORE0609 编写 OnClick() 点击事件

```java
/* 此处添加对点击事件的监听代码 */
public class HelpFragment extends Fragment implements OnClickListener {
    Activity activity;
    private RippleView iv_dache;
    private RippleView iv_daijia;
    @Override
    public View onCreateView(LayoutInflater inflater, ViewGroup container,
            Bundle savedInstanceState) {
// 获取打车代驾的 ID
View v = inflater.inflate(R.layout.fragment_help, container, false);
iv_dache = (RippleView) v.findViewById(R.id.iv_dache);
        iv_daijia = (RippleView) v.findViewById(R.id.iv_daijia);
// 设置点击事件,因为之前已经实现了 OnClickListener 接口,所以只需要参数传
this
        iv_dache.setOnClickListener(this);
        iv_daijia.setOnClickListener(this);
        return v;
    }
@Override
    public void onClick(View v) {              // 设置单击事件
        if (!ClickDouble.isFastDoubleClick()) {
            switch (v.getId()) {              // 判断所点击按钮的 id
            case R.id.iv_dache:              // 点击打车 id
                new  Thread() {
                    public void run() {
                        try {
                            sleep(300);
                        } catch (InterruptedException e) {
                            e.printStackTrace();
                        }
// 在有网的情况下跳转到打车类
startActivity(new Intent(activity,DaCheActivity.class));
    activity.overridePendingTransition(anim.fade_in,anim.fade_out);
                    };
                }.start();
                break;
            case R.id.iv_daijia:                  // 点击代驾的 id
```

项目六 打车代驾 165

```
                    new Thread() {
                        public void run() {
                            try { sleep(300);
                                } catch (InterruptedException e) {
                                    e.printStackTrace();
                            }
                        // 在有网的情况下跳转到代驾类
                        startActivity(new Intent(activity,DaiJiaActivity.class));        };
                    }.start();
                    break;
                default:
                    break;
        }  }  }
```

第五步：新建类 DaCheActivity，在 onCreate() 方法里面获取打车缓存文件。具体如代码 CORE0610 所示。

代码 CORE0610 编写获取缓存代码

```
/* 此处添加获取缓存代码 */
public class DaCheActivity extends Activity implements OnClickListener {
protected void onCreate(Bundle savedInstanceState) {
        super.onCreate(savedInstanceState);
        setContentView(R.layout.activity_dache);
        try {
        // 获取打车缓存文件
        File path = getDiskCacheDir(DaCheActivity.this, "File");
                if (!path.exists()) {
                        path.mkdirs();
                }
                mDiskLruCache = DiskLruCache.open(path, getAppVersion(Da-
CheActivity.this), 1, 10 * 1024 * 1024);
        } catch (IOException e) {
                e.printStackTrace();
        }
        findByView();
        NetWork();                          // 访问网络
    }
```

第六步：判断网络状态，有网时更新列表并显示数据，无网时，获取本地缓存信息显示在列表。具体如代码 CORE0611 所示。

代码 CORE0611 编写判断网络代码

```
/* 此处添加进行网络判断代码 */
private void NetWork() {
        if (NetUtils.isConnected(this) == true) {
            Message message = new Message();    // 当有网络时去刷新并加载
            message.what = Net_OK;
            handler.sendMessage(message);        // 给线程传递有网的信息
        } else {
            // 当没网络时直接读取本地的文件
            Message message = new Message();
            message.what = Net_NO;              // 给线程传递没有网的信息
            handler.sendMessage(message);
        } }
private Handler handler = new Handler() {
        @Override
        public void handleMessage(Message msg) {
            switch (msg.what) {
            case Net_OK:          // 当有网时，实现列表的显示并更新数据
            drivers = new ArrayList<Driver>();
driverAdapter = new DriverAdapter(DaCheActivity.this, 0, drivers, myListView);
// 上拉加载，下拉刷新列表信息
            initPullToRefreshNewsListView(myListView, driverAdapter);
            getSimulationNews();
                break;
        case Net_NO:                      // 没网时
        drivers = new ArrayList<Driver>();
driverAdapter = new DriverAdapter(DaCheActivity.this, 0, drivers, myListView);
            initPullToRefreshNewsListView(myListView, driverAdapter);
            getLocalNews();              // 获得本地缓存信息
                break;
            default:
                break;}
            super.handleMessage(msg);
        } };
```

第七步：在有网的情况下，向服务器获取信息，实现适配器实时更新。具体如代码 CORE0612 所示。

项目六 打车代驾

代码 CORE0612 编写有网情况的代码

```java
/* 此处添加在有网的情况下获取信息代码 */
// 有网的情况下
public void getSimulationNews() {
        final String url = URLCollect.BASE_URL + URLCollect.DRIVER_URL +
"page="+page+"&cate=0&pagesize=10";           // 根据路径进行数据获取
        AsyncHttpHelper.getAbsoluteUrl(url, new TextHttpResponseHandler() {
            @Override
    // 如果可以通信,则将得到的 JSON 进行解析并添加到 List 中
    public void onSuccess(int arg0, Header[] arg1, String data) {
    Gson gson = new Gson();
ArrayList<Driver> mList = gson.fromJson(data, new TypeToken<List<DRiver>
>() {
                }.getType());
                for (Driver news : mList) {
                    driverList.add(news);
                }
                driverAdapter.addNews(mList);     // 实现适配器的实时更新
                driverAdapter.notifyDataSetChanged();
                WriteToLocal(url, data);        // 将数据写入本地
            }
        @Override
        public void onFailure(int arg0, Header[] arg1, String data, Throwable arg3) {
            } }); }
```

第八步:有网时,通过 WriteToLocal() 方法将列表信息写入本地。具体如代码 CORE0613
所示。

代码 CORE0613 编写有网时将文本写入本地

```java
/* 此处添加有网的情况下将文本写入本地代码 */
private void WriteToLocal(String url, final String data) {   // 将文本写入本地
        final String key = hashKeyForDisk(url);
        new Thread(new Runnable() {
            @Override
            public void run() {
                    try {
                DiskLruCache.Editor editor = mDiskLruCache.edit(key);
                        if (editor != null) {
```

```
                              editor.set(0, data);
                              editor.commit();
                         }
                         mDiskLruCache.flush();
                } catch (IOException e) {
                         e.printStackTrace();
                } } }).start();
       }
```

第九步：无网络时，从本地获取获取缓存信息，并把信息显示在列表。具体如代码 CORE0614 所示。

代码 CORE0614 编写从本地获取文本

```
/* 此处添加没网时从本地获取文本代码 */
public ArrayList<Driver> getLocalNews() {
        String  url  =  URLCollect.BASE_URL  +  URLCollect.DRIVER_URL
"page="+page+"&cate =0&pagesize=10";              // 根据路径获取本地信息
        ArrayList<Driver> ret = new ArrayList<Driver>();
        String data = null;
        try {
                String key = hashKeyForDisk(url);
                DiskLruCache.Snapshot snapShot = mDiskLruCache.get(key);
                if (snapShot != null) {
                        data = snapShot.getString(0);
                } else {
                        data = "[]";
} } catch (IOException e) {
                e.printStackTrace();
        }
        Gson gson = new Gson();
        driverList = gson.fromJson(data, new TypeToken<List<Driver>>() {
        }.getType());
        for (Driver driver : driverList) {
                ret.add(driver);
        } return ret;
}
```

第十步：新建 DriverItemActivity 获得传参并实现打电话功能。具体如代码 CORE0615 所示。

项目六 打车代驾 169

代码 CORE0615 编写 DriverItemActivity 的内容

```java
/* 此处添加获取传递的参数代码 */
protected void onCreate(Bundle savedInstanceState) {
        super.onCreate(savedInstanceState);
        setContentView(R.layout.activity_driver);
        Intent intent = getIntent();                    // 获得跳转传参
        ID = intent.getIntExtra("id", 0);
        url = intent.getStringExtra("url");
        desc = intent.getStringExtra("desc");
        tel = intent.getStringExtra("tel");
        iv_head = (ImageView)findViewById(R.id.iv_head);
        tv_content = (TextView)findViewById(R.id.tv_content);
        iv_call = (ImageView)findViewById(R.id.iv_call);
        iv_back = (ImageButton)findViewById(R.id.iv_back);
        iv_back.setOnClickListener(this);               // 设置监听器
        iv_call.setOnClickListener(this);
        tv_content.setText(desc);                       // 接收参数
iv_head.setImageBitmap(Util.getbitmap(url));            // 通过 url 获取司机头像
    }
```

第十一步：编写点击事件，通过原生动作实现拨打电话的功能。具体如代码 CORE0616 所示。

代码 CORE0616 编写打电话功能

```java
/* 此处添加拨打电话功能的实现代码 */
@Override
    public void onClick(View v) {
        switch (v.getId()) {                            // 判断 id
        case R.id.iv_call:                              // 打电话的 id
                Intent intent = new Intent();
        // 系统默认的 action，用来打开默认的电话界面
        intent.setAction(Intent.ACTION_CALL);
        // 接收跳转传参的电话号码
                intent.setData(Uri.parse("tel:"+tel));
                DriverItemActivity.this.startActivity(intent);
                break;
        case R.id.iv_back:                              // 关闭当前页面
                DriverItemActivity.this.finish();
```

```
                        break;
            default:
                        break;
    } }
```

第十二步：新建类 DaiJiaActivity，在 onCreate() 方法里面获取缓存。具体如代码 CORE0617 所示。

代码 CORE0617 编写 DaiJiaActivity

```
/* 此处添加编写 DaiJiaActivity 代码 */
public class DaiJiaActivity extends Activity implements OnClickListener {
protected void onCreate(Bundle savedInstanceState) {
            super.onCreate(savedInstanceState);
            setContentView(R.layout.activity_dache);
            try {                          // 创建缓存
                // 获取代驾缓存文件
                        File path = getDiskCacheDir(DaiJiaActivity.this, "File");
        if (!path.exists()) {
                            path.mkdirs();
                        }
                        mDiskLruCache = DiskLruCache.open(path, getAppVersion(Dai-
JiaActivity.this), 1, 10 * 1024 * 1024);
            } catch (IOException e) {
                        e.printStackTrace();
            }
            findByView();                  // 实现界面初始化
            NetWork();                     // 访问网络
    }
```

第十三步：进行网络的判断。具体如代码 CORE0618 所示。

代码 CORE0618 编写进行网络判断的内容

```
/* 此处添加进行网络判断代码 */
private void NetWork() {
            if (NetUtils.isConnected(this) == true) {
                Message message = new Message();    // 当有网络时去刷新并加载
                message.what = Net_OK;
                handler.sendMessage(message);        // 给线程传递有网的信息
            } else {
```

项目六　打车代驾　　　　171

```java
                        // 当没网络时直接读取本地的文件
                        Message message = new Message();
                        message.what = Net_NO;
                        handler.sendMessage(message);      // 给线程传递没有网的信息
                }  }
        private Handler handler = new Handler() {
                @Override
                public void handleMessage(Message msg) {
                        switch (msg.what) {
                case Net_OK:
                        drivers = new ArrayList<Driver>();
                        // 当有网时,实现列表的显示并更新数据
                driverAdapter = new DriverAdapter(DaiJiaActivity.this, 0, drivers, myListView);
                // 上拉加载,下拉刷新列表信息
                initPullToRefreshNewsListView(myListView, driverAdapter);
                getSimulationNews();
                                break;
                case Net_NO:                          // 没网时
                        drivers = new ArrayList<Driver>();
                        driverAdapter = new DriverAdapter(DaiJiaActivity.this, 0, drivers, myList-
View);
                                initPullToRefreshNewsListView(myListView, driverAdapter);
                                getLocalNews();                // 获得本地缓存信息
                                break;
                        default:
                                break;
                        }
                super.handleMessage(msg);
        } };
```

　　第十四步:通过 getSimulationNews() 方法在有网和没网的情况下获取信息。具体如代码 CORE0619 所示。

```java
代码 CORE0619 编写有网时获取信息
/* 此处添加在有网的情况下获取信息 */
public void  getSimulationNews() {
        final String url = URLCollect.BASE_URL + URLCollect.DRIVER_URL
+"page="+page+"&cate=0&pagesize=10";          // 根据路径进行数据获取
        AsyncHttpHelper.getAbsoluteUrl(url, new TextHttpResponseHandler() {
```

```java
                @Override
        // 如果可以通信,则将得到的 JSON 进行解析并添加到 List 中
        public void onSuccess(int arg0, Header[] arg1, String data) {
        Gson gson = new Gson();
ArrayList<Driver> mList = gson.fromJson(data, new TypeToken<List<Driver>
>() {
                        }.getType());
                    for (Driver news : mList) {
                        driverList.add(news);
                    }
                    driverAdapter.addNews(mList);      // 实现适配器的实时更新
                    driverAdapter.notifyDataSetChanged();
                    WriteToLocal(url, data);          // 将数据写入本地
            }
        @Override
        public void onFailure(int arg0, Header[] arg1, String data, Throwable arg3) {
            }  });  }
```

第十五步:有网时将文本写入本地。具体如代码 CORE0620 所示。

代码 CORE0620 编写将文本写入本地

```java
/* 此处添加有网的情况下将文本写入本地 */
private void WriteToLocal(String url, final String data) {
        final String key = hashKeyForDisk(url);
        new Thread(new Runnable() {
                @Override
                public void run() {
                    try {
                    DiskLruCache.Editor editor = mDiskLruCache.edit(key);
                            if (editor != null) {
                                    editor.set(0, data);
                                    editor.commit();
                            }
                            mDiskLruCache.flush();
                    } catch (IOException e) {
                    e.printStackTrace();
                    }       }  }).start();
        }
```

项目六 打车代驾 173

第十六步：没网时从本地获取文本。具体如代码 CORE0621 所示。

代码 CORE0621 编写从本地获取文本

```
/* 此处添加没网时从本地获取文本 */
public ArrayList<Driver> getLocalNews() {
    String url = URLCollect.BASE_URL + URLCollect.DRIVER_URL +
"page="+page+"
    &cate=0&pagesize=10";                  // 根据路径获取本地信息
        ArrayList<Driver> ret = new ArrayList<Driver>();
        String data = null;
        try {
            String key = hashKeyForDisk(url);
            DiskLruCache.Snapshot snapShot = mDiskLruCache.get(key);
            if (snapShot != null) {
                data = snapShot.getString(0);
            } else {
                data = "[]";
            } } catch (IOException e) {
            e.printStackTrace();
        }
        Gson gson = new Gson();              //GSON 解析串
        driverList = gson.fromJson(data, new TypeToken<List<Driver>>() {
        }.getType());
        for (Driver driver : driverList) {
            ret.add(driver);
        } return ret;
    }
```

第十七步：新建 DriverItemActivity 获得传参并实现打电话功能。具体如代码 CORE0622 所示。

代码 CORE0622 电话服务

```
/* 此处添加获取传递的参数代码 */
protected void onCreate(Bundle savedInstanceState) {
        super.onCreate(savedInstanceState);
        setContentView(R.layout.activity_driver);
        Intent intent = getIntent();              // 获得跳转传参
        ID = intent.getIntExtra("id", 0);
        url = intent.getStringExtra("url");
```

```
                desc = intent.getStringExtra("desc");
                tel = intent.getStringExtra("tel");
                iv_head = (ImageView)findViewById(R.id.iv_head);
                tv_content = (TextView)findViewById(R.id.tv_content);
                iv_call = (ImageView)findViewById(R.id.iv_call);
                iv_back = (ImageButton)findViewById(R.id.iv_back);
                iv_back.setOnClickListener(this);          // 设置监听器
                iv_call.setOnClickListener(this);
                tv_content.setText(desc);              // 接收参数
                iv_head.setImageBitmap(Util.getbitmap(url));     // 通过 url 获取司机头像
        }
```

第十八步：调用系统复试实现拨打电话功能。具体如代码 CORE0623 所示。

代码 CORE0623 编写拨打电话功能

```
/* 此处添加拨打电话功能的实现代码 */
@Override
    public void onClick(View v) {
            switch (v.getId()) {                    // 判断 id
            case R.id.iv_call:                      // 打电话的 id
                    Intent intent = new Intent();
// 系统默认的 action，用来打开默认的电话界面
intent.setAction(Intent.ACTION_CALL);
                    intent.setData(Uri.parse("tel:"+tel));        // 接收跳转传参的电话号码
                    DriverItemActivity.this.startActivity(intent);
                    break;
            case R.id.iv_back:                      // 关闭当前页面
                    DriverItemActivity.this.finish();
                    break;
            default:
                    break;
            } }
```

第十九步：对写入本地的文本进行加密，为了保证文件的正确性，防止一些人盗用程序。此处使用到 MD5 进行加密。具体如代码 CORE0624 所示。

代码 CORE0624 编写加密文件

```
/* 此处添加对写如本地的文本进行加密代码 */
public String hashKeyForDisk(String key) {
```

```java
            String cacheKey;
            try {
                final MessageDigest mDigest = MessageDigest.getInstance("MD5");
                    mDigest.update(key.getBytes());
                    cacheKey = bytesToHexString(mDigest.digest());
            } catch (NoSuchAlgorithmException e) {
                    cacheKey = String.valueOf(key.hashCode());
            } return cacheKey;
        }
        private String bytesToHexString(byte[] bytes) {    // 使用数组将数据加密处理
            StringBuilder sb = new StringBuilder();
            for (int i = 0; i < bytes.length; i++) {
                String hex = Integer.toHexString(0xFF & bytes[i]);
                if (hex.length() == 1) {
                        sb.append('0');
                } sb.append(hex);
            } return sb.toString();
        }
```

第二十步：运行项目，实现如图 6.4 至图 6.6 效果。

本项目介绍了 U 酒保项目中打车代驾模块，了解 MD5 加密方法和 Stream 流生成方法，重点讲解 TelephonyManager 用法和电话服务的基本系统结构和服务。通过对本项目的学习可以清楚的了解电话服务机制，掌握 TelephonyManager 用法，提高对手机电话工作方式的理解。

Telephony Manager	电话管理器	device	装置，设备
traffic	通信量	state	状态
ringing	响铃	connection	连接
indicating	指示，标志	disconnected	断开的
message	信息	stream	流

一、选择题

1. 以下电话管理器的属性中表示空闲的属性是（　　）。

A.CALL_STATE_OFFHOOK　　　　　　　B.CALL_STATE_IDLE

C.CALL_STATE_RINGING　　　　　　　　D.DATA_ACTIVITY_DORMANT

2. 以下是表示数据流出的属性是（　　）。

A.DATA_ACTIVITY_INOUT　　　　　　　B.DATA_ACTIVITY_IN

C.DATA_ACTIVITY_OUT　　　　　　　　D.DATA_ACTIVITY_NONE

3. 以下关于 MD5 加密说法正确的是（　　）。

A.MD5 加密本身是可逆的,但不可破译　　B.MD5 加密本身是可逆的,也可破译

C.MD5 加密本身是不可逆的,但可破译　　D.MD5 加密本身是不可逆的,也不可破译

4. 使用一个数组的元素创建一个流的方法是（　　）。

A.Collection.stream()　　　　　　　　　B.Stream.of(T...)

C.Stream.of(T[])　　　　　　　　　　　D.Stream.empty()

5. 对 toArray() 的描述正确的是（　　）。

A. 将流的元素聚合为一个汇总值

B. 将流的元素聚合到一个汇总结果容器中

C. 通过比较符返回流的最大元素

D. 使用流的元素创建一个数组

二、填空题

1. 一个电话的基本功能有:拨叫电话、_____、挂断电话、_____、网络连接和_____。

2. MD5 加密的特点:_____、_____、_____、_____。

3. 中间操作负责将一个流转换为另一个流,中间操作包括_____、_____、_____、limit() 和_____。

4. TelephonyManager 的功能是:_____,_____,侦听电话状态以及_____。

5. MD5 算法分析:MD5 以_____位分组来处理输入的信息,每一分组被划分为_____个 32 位子分组,经过了一系列处理后,算法输出由四个_____位分组组成,将这四个位分组级联后将生成一个_____位散列值。

三、上机题

1. 编写代码实现打电话功能。

2. 编写代码实现 MD5 加密功能。

项目七 无忧险

通过 U 酒保项目无忧险模块的实现,了解不同 UI 控件之间的区别,掌握无忧险界面更新的方法,学习上拉加载下拉刷新保险信息列表,具有编写刷新加载信息列表的能力。在任务实现过程中:

- 了解不同 UI 控件之间的区别。
- 掌握无忧险界面更新的方法。
- 掌握上拉加载下拉刷新保险信息列表。

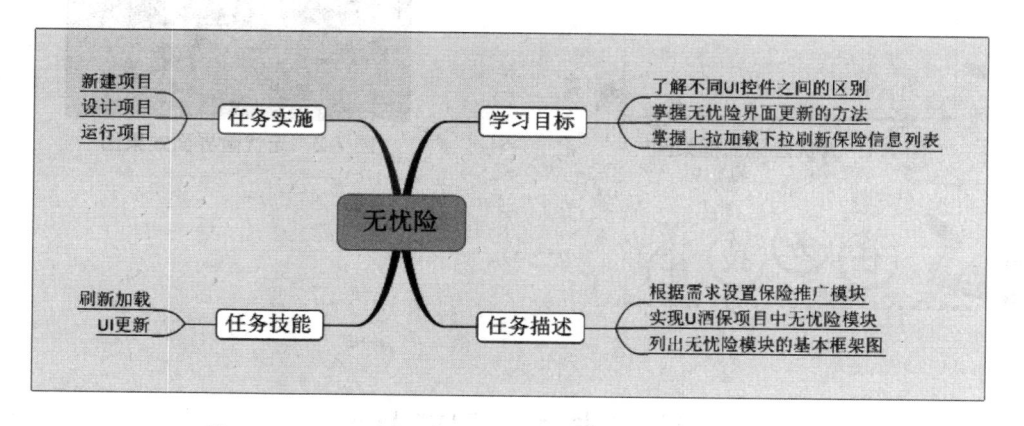

【情境导入】

U 酒保研发团队为保证用户的人身安全和利益,设置了保险推广模块。该模块设有多家保险公司可供用户选择,用户可根据自身的情况为人与车购买有效合理的保险。合理购买保险,最大程度地减少用户在发生意外时的经济损失,从根本上解决用户的实际问题。本模块通过无忧险模块的实现,讲解了如何实时更新保险信息并将其显示到界面。

【功能描述】

本项目将实现 U 酒保项目中无忧险模块:

● 实现下拉刷新的功能。
● 实现上拉加载的功能。
● 实现 UI 更新的功能。

【基本框架】

基本框架如图 7.1 所示。通过本模块的学习,能将框架图 7.1 转换成效果图 7.2 所示。

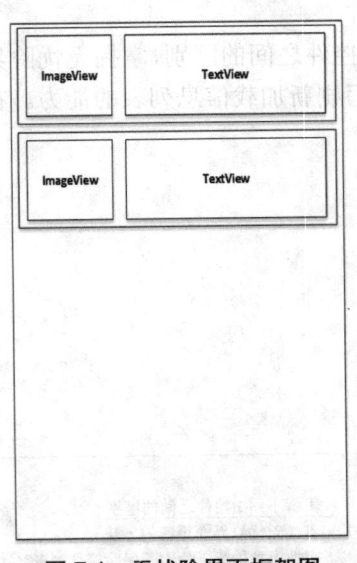

图 7.1　无忧险界面框架图

图 7.2　无忧险界面效果图

技能点 1　刷新加载

1　PullToRefresh(刷新)简介

PullToRefresh 是一个强大的拉动刷新库,用来实现多种控件的刷新操作。如 ListView、ViewPager(多页显示控件)、WebView(网络视图)、ExpandableListView(实现下拉 list)、Grid-View(多控件布局)、(Horizontal)ScrollView(循环滚动控件)、Fragment 等。在开发过程中,

项目七　无忧险　　　179

PullToRefresh 库的实现机制如图 7.3 所示。

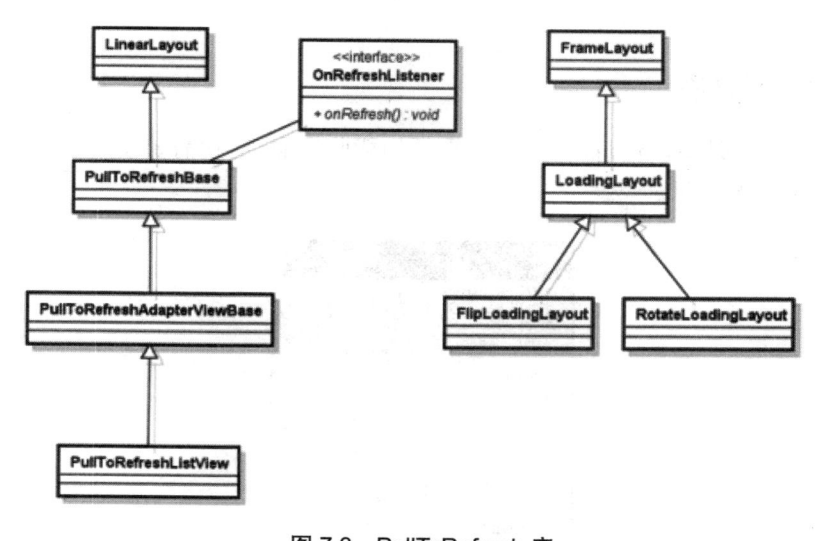

图 7.3　PullToRefresh 库

　　PullToRefreshBase 作为父类，扩展了 LinearLayout 水平布局，如果使用 ListView 则需要观看子类 PullToRefreshAdapterViewBase 和 PullToRefreshListView。PullToRefreshAdapterView-Base 子类中提供了 isHeaderShown() 和 isFooterShown() 两个接口，用来选择是向上刷新方式还是向下刷新方式。PullToRefreshListView 子类中构造刷新的 ListView 控件，设置监听对象，在刷新时更新顶部或是底部的刷新标签。

　　下拉刷新主要由 PullToRefresh 库实现的，下拉刷新的父 View 是 LinearLayout，在 Linear-Layout 中添加了 HeaderView（头部视图）、FooterView（底部视图）和 ListView。如图 7.4 所示。

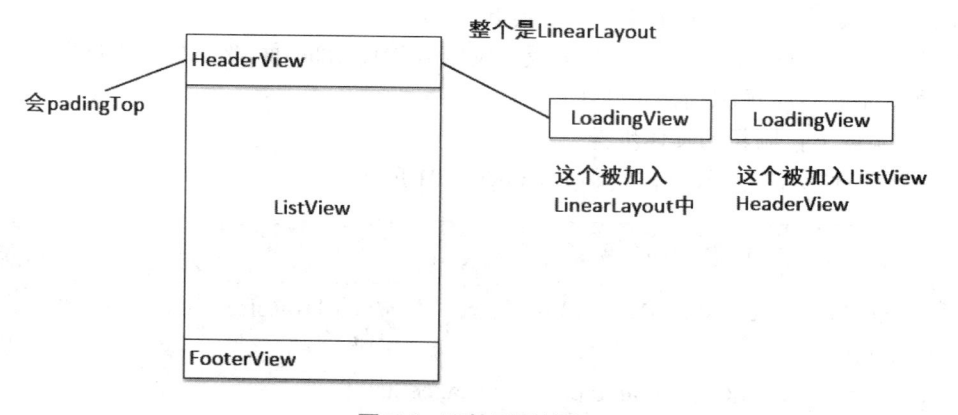

图 7.4　下拉刷新结构

　　图 7.4 所示的第一个 LoadingLayout 主要显示下拉刷新时的文字"释放开始刷新"；第二个 LoadingLayout 显示松手后的文字"正在刷新"。

● HeaderView：下拉刷新时露出的上面部分，下拉到一定位置，松手后会开始请求网络数据,然后刷新 ListView 的列表。

● FooterView:是 ListView 手势一直上滑到显示出最后一条数据,然后继续按住滑动到一定位置,再松手,会加载下一页的数据。

2 实现下拉刷新方法

当前界面信息不是最新信息时,就需要重新加载此界面,这就需要使用下拉刷新来实现界面更新,效果如图 7.5 所示。

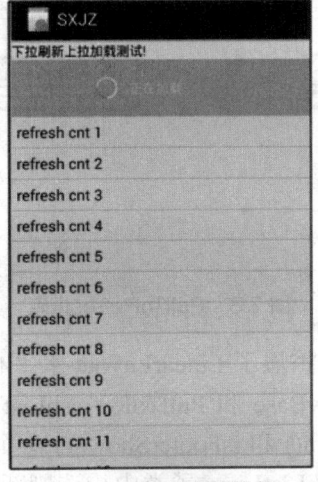

图 7.5 下拉刷新效果

实现下拉刷新主要的流程是:

● 下拉,显示提示头部界面(HeaderView),这个过程提示用户"下拉刷新"。

● 下拉到一定程度,超出了刷新最基本的下拉界限,系统认为达到了刷新的条件,提示用户"可以松手刷新了",效果上允许用户继续下拉。

● 用户下拉后提示头部界面,所以先反弹仅显示提示头部界面,然后提示用户"正在加载"。

● 刷新完成后,隐藏提示头部界面。

以下为实现下拉刷新的具体步骤:

(1)设置控件所需布局,具体代码如 CORE0701 所示。

代码 CORE0701 设置控件所需布局

```
private PullToRefreshListView pListView;        // PullToRefreshListView 控件对象
@Override
 protected void onCreate(Bundle savedInstanceState) {
    super.onCreate(savedInstanceState);
    setContentView(R.layout.listview_layout);
    pListView = (PullToRefreshListView) findViewById(R.id.plistview);
 }
```

项目七 无忧险 181

（2）为 ListView 绑定适配器，具体代码如 CORE0702 所示。

代码 CORE0702 为 ListView 绑定适配器

```
ArrayList<String> arrayList = new ArrayList<String>();
 adapter = new ArrayAdapter<String>(this, R.layout.item_layout, R.id.tv_item_name,
arrayList);
// 初始化适配器
adapter.add("snail");
adapter.add(" _snail");
adapter.add(" __snail");
adapter.add("___snail");
pListView.setAdapter(adapter);                // 绑定适配器
```

（3）编写控件布局，具体代码如 CORE0703 所示。

代码 CORE0703 编写控件布局

```
<com.handmark.pulltorefresh.library.PullToRefreshListView
      android:id="@+id/plistview"
android:layout_width="match_parent"
android:layout_height="match_parent"
/>
...
```

（4）设置界面刷新模式，具体代码如 CORE0704 所示。

代码 CORE0704 设置刷新模式

```
/*
 * 设置 PullToRefresh 刷新模式
 * BOTH: 上拉刷新和下拉刷新都支持
 * DISABLED: 禁用上拉下拉刷新
 * PULL_FROM_START: 仅支持下拉刷新（默认）
 * PULL_FROM_END: 仅支持上拉刷新
 * MANUAL_REFRESH_ONLY: 只允许手动触发
 *
 */
pListView.setMode(Mode.PULL_FROM_START);
```

（5）绑定刷新监听事件，具体代码如 CORE0705 所示。

代码 CORE0705 绑定刷新监听事件

```
// 设置刷新监听
pListView.setOnRefreshListener(new OnRefreshListener<ListView>() {
    @Override
    public void onRefresh(PullToRefreshBase<ListView> refreshView) {
    String str = DateUtils.formatDateTime(MainActivity.this, System.
    currentTimeMillis(),DateUtils.FORMAT_SHOW_TIME | DateUtils.
    FORMAT_SHOW_DATE | DateUtils.FORMAT_ABBREV_ALL);
// 设置刷新标签
pListView.getLoadingLayoutProxy().setRefreshingLabel(" 正在加载 ");
// 设置下拉标签
pListView.getLoadingLayoutProxy().setPullLabel(" 下拉刷新 ");
// 设置释放标签
pListView.getLoadingLayoutProxy().setReleaseLabel(" 松开刷新数据 ");
// 设置上一次刷新的提示标签
refreshView.getLoadingLayoutProxy().setLastUpdatedLabel(" 最后更新时间 :" + str);
new MyTask().execute();                // 加载数据操作
    }   });
```

3 实现上拉加载方法

当用户从网络上读取空间内容的时候,如果立刻加载用户未读的全部内容,将耗费较长的时间,使得用户体验较差,同时整屏的大小也不能够显示全部的内容。所以需要用到另一个功能,那就是 ListView 的分页,也就是上拉加载,用户可根据需求加载数据。

上拉加载的方法和下拉刷新的方法基本相同,不同点主要体现在设置刷新模式和绑定监听事件。实现效果如图 7.6 所示。

(1)设置刷新模式,具体代码如 CORE0706 所示。

代码 CORE0706 设置刷新模式

```
// 判断头部是否展示出来,如果展示出来代表下拉使得头部展示。true 为下拉
public boolean isShownHeader() {
    return getHeaderLayout().isShown();
}
// 判断底部是否展示出来,如果展示出来代表上拉使得底部展示。true 为上拉
public boolean isShownFooter() {
    return getFooterLayout().isShown();
}
```

项目七　无忧险　　183

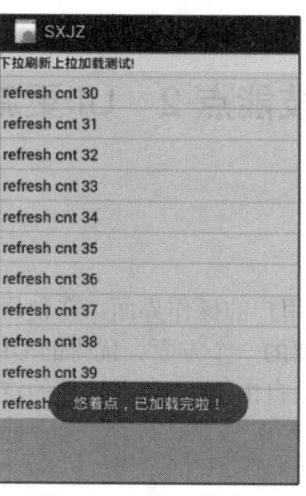

图 7.6　上拉加载效果

（2）绑定监听事件，具体代码如 CORE0707 所示。

```
代码 CORE0707  绑定监听事件
    // 设置刷新监听
    pListView.setOnRefreshListener(new OnRefreshListener<ListView>() {
      @Override
      public void onRefresh(PullToRefreshBase<ListView> refreshView) {
        String str = DateUtils.formatDateTime(MainActivity.this, System.currentTimeMi-
llis(), DateUtils.FORMAT_SHOW_TIME | DateUtils.FORMAT_SHOW_DATE | DateUtils.
FORMAT_ABBREV_ALL);
        // 上拉加载更多业务代码
        if(refreshView.isShownFooter()) {
          pListView.getLoadingLayoutProxy().setRefreshingLabel(" 正在加载 ");
          pListView.getLoadingLayoutProxy().setPullLabel(" 上拉加载更多 ");
          pListView.getLoadingLayoutProxy().setReleaseLabel(" 释放开始加载 ");
          refreshView.getLoadingLayoutProxy().setLastUpdatedLabel(" 最后加载时
间 :" + str);
          new MyTask().execute();
        } } });
```

　　拓展　下拉刷新几乎是每个 Android 应用需要的功能。前边已经对 Pull-
ToRefresh 进行了学习，要清楚探索技能的道路是没有尽头的，一个强大的 Android 下拉刷新框架——Ultra-Pull-To-Refresh（简称 UltraPTR）值得我们去探
索。这不，下方的二维码已经为你指明方向。还不赶快行动？

技能点 2　UI 更新

1　UI 简介

UI 是用户界面的简称。泛指用户的操作界面。在使用上，对软件的人机交互、操作逻辑、界面美观的整体设计则是同样重要的一个方面。优秀的 UI 不仅能让软件变得有个性有品位，还能让软件的操作变得舒适、简单、自由，充分体现软件的定位和特点。

2　UI 设计

UI 设计主要指界面的样式，美观程度。软件界面设计就像工业产品中的工业造型设计一样，是产品的重点。一个美观的界面会给人带来舒适的视觉享受，拉近人与电脑的距离。界面设计不是单纯的美术绘画，它需要定位使用者、使用环境、使用方式并且为用户而设计，是纯粹的科学性艺术设计。检验一个界面的标准既不是某个项目开发组领导的意见也不是项目成员投票的结果，而是最终用户的感受。所以界面设计要和用户研究紧密结合，是一个不断为最终用户设计满意视觉效果的过程。

3　UI 控件

对于日益增加的 UI 控件需求，市场上也出现了很多可供选择的 UI 控件，满足用户比较复杂的需求。这些控件帮助简化 UI 设计工作，提高效率。在日常使用的应用程序当中，凡是显示在屏幕上的"可视化"控件都是 View。图 7.7 展示了 Android 中的 View 元素体系。

UI 控件可以分为以下几类：

● Android UI 控件：文本控件、按钮控件、状态开关控件、单选与复选按钮、图片控件、时钟控件、日期与时间选择控件等。

● Web UI 控件：图表和图形、日期和日历、组合框、对话框、进度条、布局和翻译编辑器等。

● iOS 基本 UI 控件：Button 控件、开关控件、滑块控件、工具栏、WebView 等。

UI 控件的三要素：绘制、数据、控制。

● 绘制：在界面中可见的图形，每一个控件都有自己的样式，如 TableView 是一张数据表，又如 datePicker 是一个时间选择器，它们的样式是不同的。

● 数据：控件也需要自己的数据，如 label，需要显示文字的数据，如 ImageView 需要显示图片的数据，如果没有数据这些控件的使用将会变得没有意义。

● 控制：最典型的就是 Button，这是用户与界面交互的关键，还有其他的控件，如 Scrollview，可以滑动加载数据。

4　UI 更新的三种方法

在 Android 项目中，子线程中完成耗时操作之后要更新 UI，本部分详细讲解更新 UI 的三种方法。

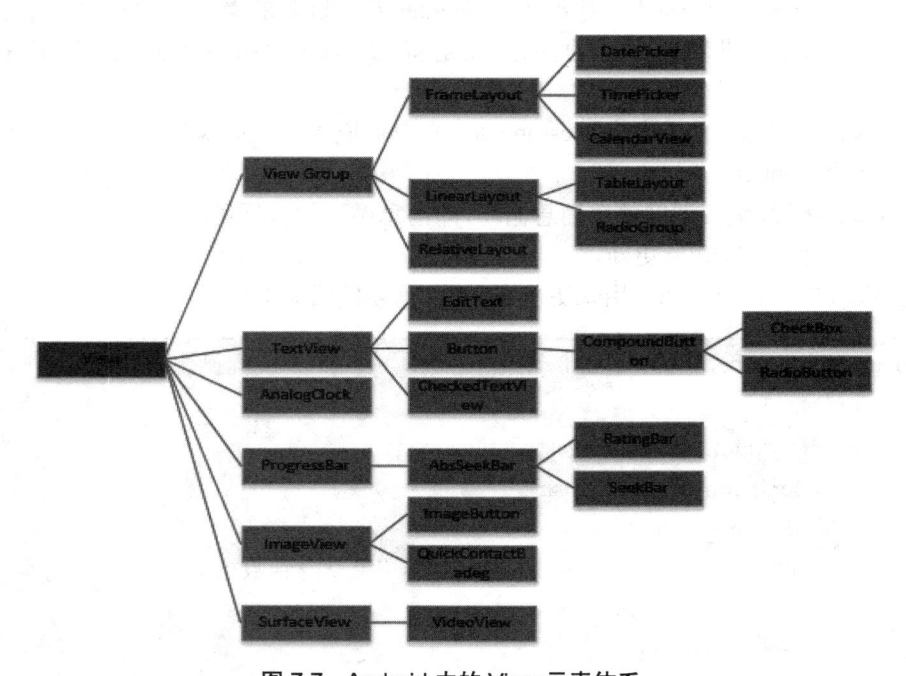

图 7.7　Android 中的 View 元素体系

首先需要了解 Android 中的消息机制，如图 7.8 所示。

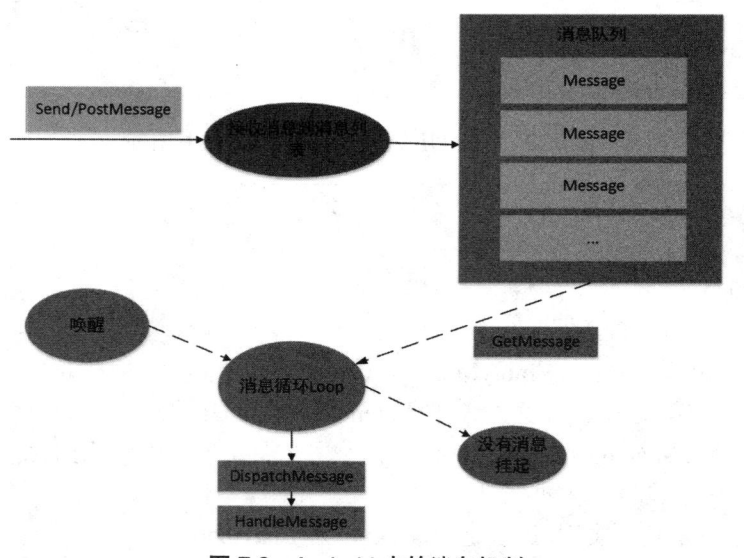

图 7.8　Android 中的消息机制

➤ Message：消息，其中包含了消息 ID，消息处理对象以及处理的数据等，由 Message-Queue 统一列队，终由 Handler 处理。

➤ Handler：处理者，负责 Message 的发送及处理。使用 Handler 时，需要实现 handleMessage(Message msg) 方法来对特定的 Message 进行处理，例如更新 UI 等。

➢ MessageQueue：消息队列，用来存放 Handler 发送过来的消息，并按照 FIFO 规则执行。当然，存放 Message 并非实际意义的保存，而是将 Message 以链表的方式串联起来的，等待 Looper 的抽取。

➢ Looper：消息泵，不断地从 MessageQueue 中抽取 Message 执行。因此，一个 MessageQueue 需要一个 Looper。

➢ Thread：线程，负责调度整个消息循环，即消息循环的执行场所。

● 方法一：使用 Handler 更新。

（1）主线程中定义 Handler，用来更新主界面，具体代码如下所示。

```java
Handler mHandler = new Handler() {          //handler 线程抛出处理数据
    @Override
    public void handleMessage(Message msg) {
        super.handleMessage(msg);
        switch (msg.what) {
        case 0:                    // 完成主界面更新，得到数据
            String data = (String)msg.obj;
            updateWeather();
            textView.setText(data);
            break;
        default:
            break;
        } } };
```

（2）线程发消息，通过 Handler 完成 UI 更新，具体代码如下所示。

```java
private void updateWeather() {
    new Thread(new Runnable(){
        @Override
        public void run() {
            mHandler.sendEmptyMessage(0);
        // 耗时操作，完成之后发送消息给 Handler，完成 UI 更新
            Message msg =new Message();          // 需要数据传递，用下面方法
            msg.obj = " 数据 ";
        // 数据可以是基本类型，可以是对象，可以是 List、map 等
            mHandler.sendMessage(msg);
        }
    }).start();
}
```

项目七　无忧险　　187

该方法的 Handler 对象必须定义在主线程中,如果是多个类直接互相调用,就需要传递 content 对象或通过接口调用。

● 方法二:用 Activity 对象的 runOnUiThread() 方法更新。

在子线程中通过 runOnUiThread() 方法更新 UI,具体代码如下所示。

```
new Thread() {
    public void run() {
        runOnUiThread(new Runnable(){        // 这是耗时操作,完成之后更新 UI
            @Override
            public void run() {
                imageView.setImageBitmap(bitmap);        // 更新 UI
            } }); }  }.start();
```

如果在非上下文类中(Activity),可以通过传递上下文实现调用。

● 方法三:View.post(Runnable) 更新 UI。

```
imageView.post(new Runnable(){
        @Override
        public void run() {
            imageView.setImageBitmap(bitmap);
        }   });
```

该方法简单,但是需要传递要更新 View。

拓展　通过学习已经知道 Android 更新 UI 的方法。但是在开发程序的过程中难免会遇到错误,作者已经列出了最容易遇到的一个错误异常,俗话说得好上有政策下有对策,想要一探究竟,扫码便知。

通过以下步骤实现如图 7.2 所示 U 酒保的无忧险模块。

具体步骤如下所示。

第一步：将 UJB_01 导入工程，在其基础上进一步实现 UJB 项目无忧险模块。首先点击 "Open an existing Android Studio project" 打开磁盘路径查找所需项目并导入，如图 7.9 和图 7.10 所示。实现如图 7.11 所示结果图。

图 7.9 导入项目

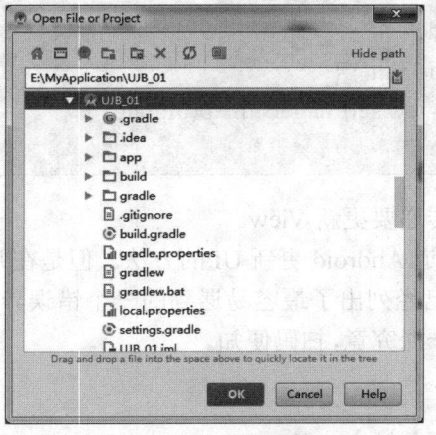

图 7.10 工程目录

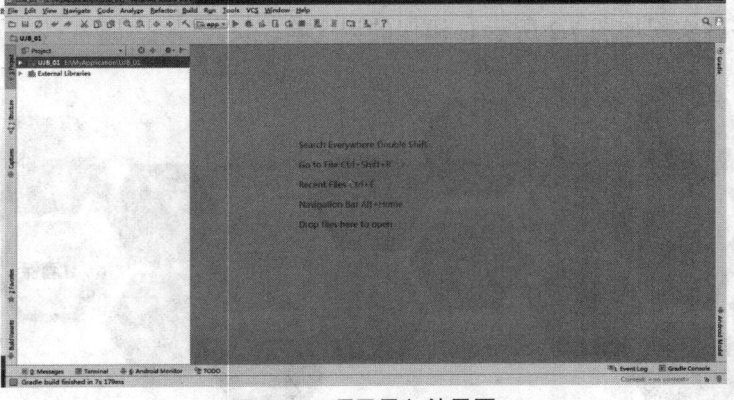

图 7.11 项目导入结果图

项目七 无忧险 189

第二步：在 src 文件夹下建立 MainActivity.java 文件中，实现点击"无忧险"小标，跳转到 AIFragment。具体如代码 CORE0708 所示。

代码 CORE0708 编写跳转到 AIFragment

```
/* 此处添加点击无忧险图标实现跳转到 AIFragment 代码 */
public class MainActivity extends FragmentActivity implements OnClickListener {
public static final String TAG_1 = "HOME";        // 给每个 TabHost 设置标签
public static final String TAG_2 = "ADVISE";
public static final String TAG_3 = "HELP";
public static final String TAG_4 = "INSURANCE";
@Override
protected void onCreate(Bundle savedInstanceState) {
super.onCreate(savedInstanceState);
setContentView(R.layout.activity_tabs);
PushAgent mPushAgent = PushAgent.getInstance(this.getApplication());
mPushAgent.enable();
mTabHost = (TabHost) findViewById(android.R.id.tabhost);
mTabHost.setup();
mTabManager = new TabManager(this, mTabHost, R.id.realtabcontent);
mTabanager.addTab(mTabHost.newTabSpec(TAG_1).setIndicator(createTabView(" 安
全自测 ", R.drawable.tabbar_home_selector)), HomeFragment.class, null);
mTabManager.addTab(mTabHost.newTabSpec(TAG_2).setIndicator(createTabView("
健康助手 ", R.drawable.tabbar_msg_selector)), AdviseFragment.class, null);
    mTabManager.addTab(mTabHost.newTabSpec(TAG_3).setIndicator(createTabView("
打车 / 代驾 ", R.drawable.tabbar_profile_selector)), HelpFragment.class, null);
    mTabManager.addTab(mTabHost.newTabSpec(TAG_4).setIndicator(createTabView("
无忧险 ", R.drawable.tabbar_discover_selector)), AIFragment.class, null);
```

第三步：在实现上拉加载下拉刷新功能之前要导入 library_pullToRefresh.jar，具体步骤如下所示。

（1）复制 library_pullToRefresh.jar 包，添加到如图 7.12 所示目录下。

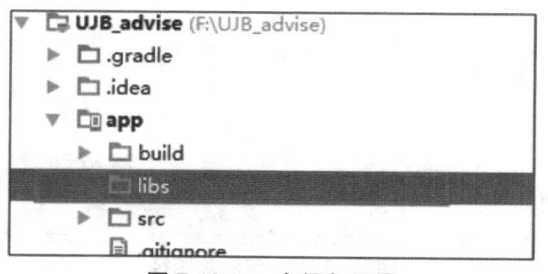

图 7.12 jar 包添加目录

（2）选择该项目，右击鼠标出现如图 7.13 所示界面，点击"Open Module Settings"，跳转到下一界面，选择"app"→"Dependencise"，如图 7.14 所示。

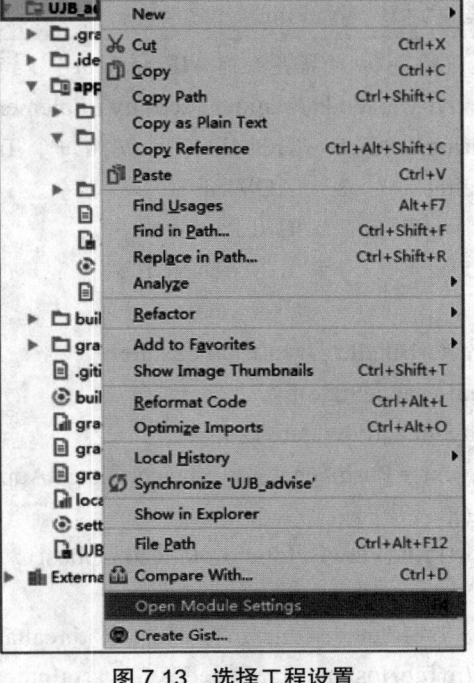

图 7.13　选择工程设置

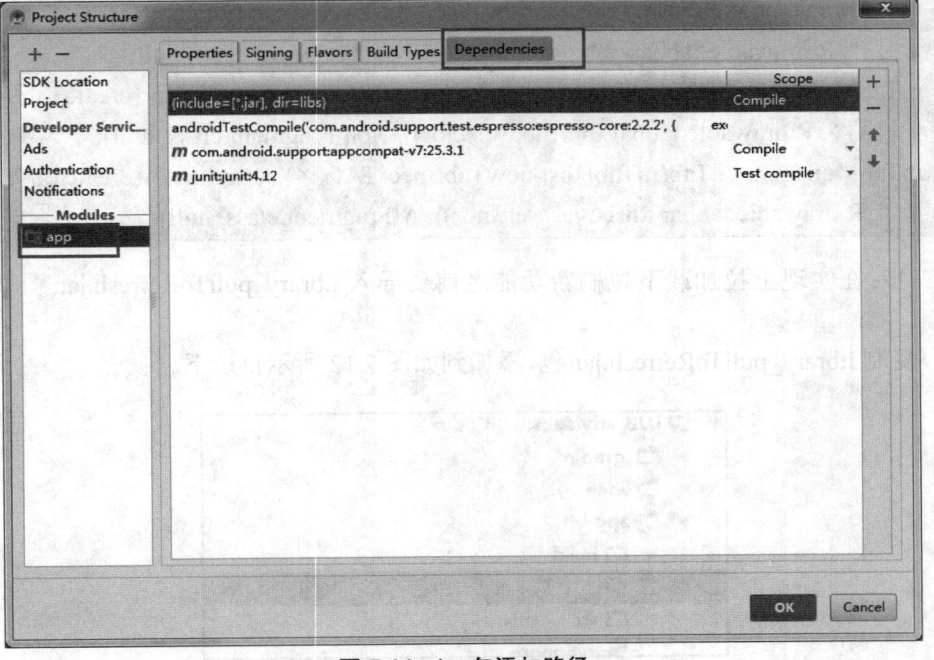

图 7.14　jar 包添加路径

（3）单击"+"显示选择栏，如图 7.15 所示。选择 File dependency，跳转到下一界面，如图 7.16 所示，选择添加的 jar 包后点击"OK"。

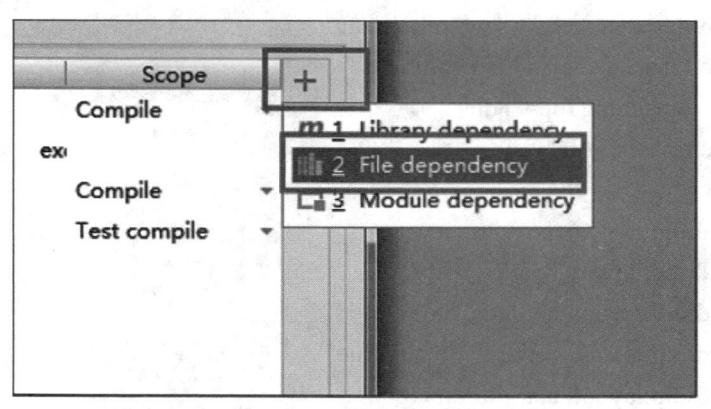

图 7.15　选择文件属性

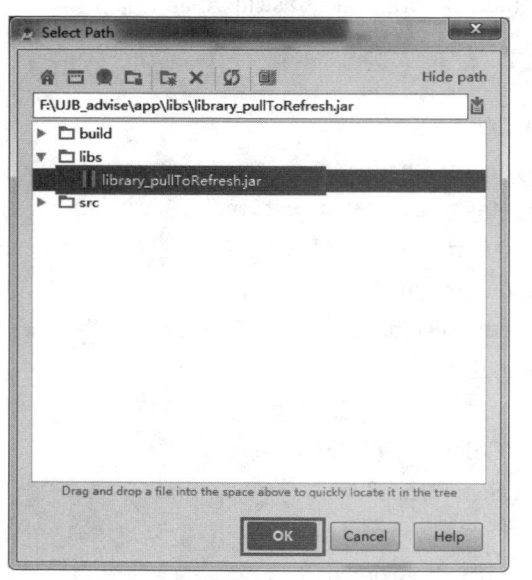

图 7.16　选择需要添加的 jar 包

第四步：对跳转到的界面也就是无忧险模块的布局文件进行设计。具体如代码 CORE0709 所示。

代码 CORE0709　编写跳转到界面布局代码

```
/* 此处添加无忧险界面布局代码 */
<FrameLayout xmlns:android="http://schemas.android.com/apk/res/android"
    android:layout_width="match_parent"
    android:layout_height="wrap_content" >
<RelativeLayout
```

```xml
    android:id="@+id/rl_typ1"
    android:layout_width="match_parent"
    android:layout_height="wrap_content"
    android:background="#303030"
    android:paddingTop="8dp"
    android:visibility="visible" >
<ImageView
        android:id="@+id/ivPreview"
        android:layout_width="94dp"
        android:layout_height="70dp"
        android:layout_centerVertical="true"
        android:layout_marginLeft="8dp"
        android:layout_marginRight="10dp"
        android:contentDescription="@string/blank"
        android:scaleType="fitXY" />
<TextView
        android:id="@+id/tvTitle"
        android:layout_width="wrap_content"
        android:layout_height="wrap_content"
        android:layout_marginLeft="13.5dp"
        android:layout_marginTop="8dp"
        android:layout_toRightOf="@id/ivPreview"
        android:singleLine="true"
        android:text=""
        android:textColor="#FFFFFF"
        android:textSize="15sp" />
<TextView
        android:id="@+id/tvContent"
        android:layout_width="wrap_content"
        android:layout_height="wrap_content"
        android:layout_alignLeft="@id/tvTitle"
        android:layout_below="@id/tvTitle"
        android:layout_marginLeft="2dp"
        android:layout_marginTop="8dp"
        android:maxLines="2"
        android:text=""
        android:textColor="#FFFFFF"
        android:textSize="12sp" />
```

```xml
<View android:id="@+id/line"
        android:layout_width="match_parent"
        android:layout_height="0.5dp"
        android:layout_marginTop="16dp"
        android:layout_below="@+id/tvContent"
        android:background="@drawable/fenge"/>
</RelativeLayout>
<RelativeLayout
    android:id="@+id/rl_typ2"
    android:layout_width="match_parent"
    android:layout_height="wrap_content"
    android:background="#303030"
    android:padding="4dp"
    android:visibility="gone" >
<TextView
        android:id="@+id/tvTitle_2"
        android:layout_width="wrap_content"
        android:layout_height="wrap_content"
        android:layout_marginBottom="4dp"
        android:layout_marginLeft="10dp"
        android:layout_marginTop="1dp"
        android:singleLine="false"
        android:text=""
        android:textColor="#FFFFFF"
        android:textSize="15sp" />
<LinearLayout
        android:layout_width="match_parent"
        android:layout_height="wrap_content"
        android:layout_below="@+id/tvTitle_2"
        android:orientation="horizontal" >
<ImageView
        android:id="@+id/iv_pic1"
        android:layout_width="94dp"
        android:layout_height="70dp"
        android:layout_marginLeft="8dp"
        android:layout_weight="1"
        android:contentDescription="@string/blank"
        android:scaleType="fitXY" />
```

```
    <ImageView
            android:id="@+id/iv_pic2"
            android:layout_width="94dp"
            android:layout_height="70dp"
            android:layout_marginLeft="8dp"
            android:layout_weight="1"
            android:contentDescription="@string/blank"
            android:scaleType="fitXY" />
    <ImageView
            android:id="@+id/iv_pic3"
            android:layout_width="94dp"
            android:layout_height="70dp"
            android:layout_marginLeft="8dp"
            android:layout_weight="1"
            android:contentDescription="@string/blank"
            android:scaleType="fitXY" />
</LinearLayout> </RelativeLayout> </FrameLayout>
```

第五步：在进入无忧险界面后首先要判断网络是否连接，并给线程传递相应的信息。具体如代码 CORE0710 所示。

```
代码 CORE0710 编写判断网络的方法
/* 此处添加判断网络的代码 */
    private void NetWork() {
        // 判断网络是否已经连接
        if (NetUtils.isConnected(activity) == true) {
            Message message = new Message();       // 当有网络时去刷新并加载
            message.what = Net_OK;
handler.sendMessage(message);          // 给线程传递有网的信息
        } else {
            Message message = new Message();       // 当没网络时直接读取本地
的文件
            message.what = Net_NO;
handler.sendMessage(message);          // 给线程传递没网的信息
        } }
```

第六步：在有网的情况下线程接收消息，适配数据并绑定适配器。具体如代码 CORE0711 所示。

项目七 无忧险 195

代码 CORE0711 编写有网情况下的代码

```java
/* 此处添加有网情况下的代码 */
// 使用 handler 接受消息
Handler handler = new Handler() {
        @Override
        public void handleMessage(Message msg) {
                switch (msg.what) {
                case Net_OK:                    // 网络已经连接
                        detail_loading.setVisibility(View.GONE); // 隐藏加载效果
                        insurs = new ArrayList<Insur>();        // 创建 ArrayList 的对象
                        mAdapter = new NewsAdapter(activity, 0, insurs, mListView);
// 给适配器适配数据
initPullToRefreshNewsListView(mListView, mAdapter); // 绑定适配器
    getSimulationNews();
                        break;
                case Net_NO:                    // 网络没有连接
detail_loading.setVisibility(View.GONE);          // 隐藏加载效果
                        insurs = new ArrayList<Insur>();     // 创建 ArrayList 对象
    mAdapter = new NewsAdapter(activity, 0, insurs, mListView); // 创建适配器对象并
传递参数
    initPullToRefreshNewsListView(mListView, mAdapter); // 加载数据并适配
    getLocalNews();                         // 获得本地缓存的方法
                        break;
                default:
                        break;
                }super.handleMessage(msg);
        }
```

第七步: 创建 NewsAdapter 适配器(继承 ArrayAdapter<Insur>), 重写 NewsAdapter 方法, 并传递参数。具体如代码 CORE0712 所示。

代码 CORE0712 编写 NewsAdapter 适配器

```java
/* 此处添加适配器代码 */
public NewsAdapter(Context context, int resource, List<Insur> objects,PullToRefreshListView
    listView) {
            super(context, resource, objects);
            this.pullToRefreshListView = listView;
```

```java
        this.insurs = objects;
// 创建 HashSet 对象
taskCollection = new HashSet<BitmapWorkerTask>();
int maxMemory = (int) Runtime.getRuntime().maxMemory();
// 获取应用程序最大可用内存
int cacheSize = maxMemory / 8;
// 设置图片缓存大小为程序最大可用内存的 1/8
        mMemoryCache = new LruCache<String, Bitmap>(cacheSize) {
            @Override
            protected int sizeOf(String key, Bitmap bitmap) {
                return bitmap.getByteCount();
            } };
        try {
        // 获取图片缓存路径
        File cacheDir = getDiskCacheDir(context, "thumb");
        if (!cacheDir.exists()) {
                cacheDir.mkdirs();
            }
        // 创建 DiskLruCache 实例，初始化缓存数据
        mDiskLruCache = DiskLruCache.open(cacheDir,
         getAppVersion(context),1, 10 * 1024 * 1024);
        } catch (IOException e) {
            e.printStackTrace();
        } }
    @Override
    public View getView(int position, View convertView, ViewGroup parent) {
        final Insur insur = getItem(position);
        String[] sourceStrArray = null;        // 最多分割出 3 个字符串
        ViewHolder holder;
        if (convertView == null) {
        // 判断 View 视图是否为空。判断是否第一次进入
        convertView = LayoutInflater.from(getContext()).inflate(
                    R.layout.item_news, null);
            holder = new ViewHolder();        // 初始化控件
holder.rl_typ1 = (RelativeLayout) convertView            .findViewById(R.id.rl_typ1);
holder.rl_typ2 = (RelativeLayout) convertView.findViewById(R.id.rl_typ2);
holder.ivPreview = (ImageView) convertView.findViewById(R.id.ivPreview);
holder.tvTitle = (TextView) convertView.findViewById(R.id.tvTitle);
```

```java
                holder.tvContent = (TextView) convertView.findViewById(R.id.tvContent);
                holder.tvTitle_2 = (TextView) convertView.findViewById(R.id.tvTitle_2);
                holder.iv_pic1 = (ImageView) convertView.findViewById(R.id.iv_pic1);
                holder.iv_pic2 = (ImageView) convertView.findViewById(R.id.iv_pic2);
                holder.iv_pic3 = (ImageView) convertView.findViewById(R.id.iv_pic3);
                convertView.setTag(holder);                 // 设置一个标签
            } else {
                // 如果不为空，获得 Tag 标签
                holder = (ViewHolder) convertView.getTag();
            }
            String url = insur.getInsur_pic();              // 对图片的处理
            if (!"pic".equals(insur.getInsur_desc())) {
                holder.rl_typ1.setVisibility(View.VISIBLE);
                holder.rl_typ2.setVisibility(View.GONE);
                holder.ivPreview.setTag(insur.getInsur_pic())     // 设置 Tag 标签
                // 设置图片资源
                holder.ivPreview.setImageResource(R.drawable.detail_loading);
                loadBitmaps(holder.ivPreview, url);
                holder.tvTitle.setText(insur.getInsur_name());        // 设置标题内容
                holder.tvContent.setText(insur.getInsur_desc());
            } else {
                holder.rl_typ1.setVisibility(View.GONE);      // 设置为隐藏
                holder.rl_typ2.setVisibility(View.VISIBLE);   // 设置为不可见
                holder.tvTitle_2.setText(insur.getInsur_name()); // 设置 title 内容
                // 根据路径获得加载内容
                sourceStrArray = url.split("\\|");
                holder.iv_pic1.setTag(sourceStrArray[0]);
                loadBitmaps(holder.ivPreview, sourceStrArray[0]);
                holder.iv_pic2.setTag(sourceStrArray[1]);
                loadBitmaps(holder.ivPreview, sourceStrArray[1]);
                holder.iv_pic3.setTag(sourceStrArray[2]);
                loadBitmaps(holder.ivPreview, sourceStrArray[2]);
            } return convertView;
        }
        // 封装保险信息并添加
        public void addNews(List<Insur> addNews) {
            for (Insur hm : addNews) {
                insurs.add(hm);
```

```
        }notifyDataSetChanged();
    }
    static class {                    // 自定义 ViewHolder 类
        RelativeLayout rl_typ1;
        ImageView ivPreview;
        TextView tvTitle;
        TextView tvContent;
        RelativeLayout rl_typ2;
        TextView tvTitle_2;
        ImageView iv_pic1;
        ImageView iv_pic2;
        ImageView iv_pic3;
    }
```

第八步：通过 LvOnRefreshListener(ptrlv) 方法实现下拉刷新上拉加载，并将数据绑定在列表。具体如代码 CORE0713 所示。

代码 CORE0713 编写绑定适配器代码

```
/* 此处添加绑定适配器的代码 */
// 将传入的 PullToRefreshListView ptrlv, NewsAdapter adapter 绑定在列表
    private void initPullToRefreshNewsListView(PullToRefreshListView ptrlv, News-
Adapter adapter) {
        ptrlv.setMode(Mode.BOTH);        // 设置模式为同时支持上拉加载和下拉刷新
        ptrlv.setOnRefreshListener(new LvOnRefreshListener(ptrlv));    // 设置刷新监听器
        ptrlv.setAdapter(adapter);       // 绑定适配器
    }
// 通过 LvOnRefreshListener(ptrlv) 方法实现下拉刷新上拉加载
    class LvOnRefreshListener implements OnRefreshListener2<ListView> {
        private PullToRefreshListView mPtflv;
        public LvOnRefreshListener(PullToRefreshListView ptflv) {
            this.mPtflv = ptflv;
        }
```

第九步：实现下拉刷新上拉加载的功能。具体如代码 CORE0714 所示。

代码 CORE0714 编写下拉刷新上拉加载

```
/* 此处添加下拉刷新上拉加载的代码 */
        // 通过 onPullDownToRefresh() 方法实现下拉刷新
        @Override
```

```java
        public void onPullDownToRefresh(PullToRefreshBase<ListView> refresh-
View) {
        // 格式化时间，设置时间显示样式
        String label = DateUtils.formatDateTime(activity.getApplicationContext(), System.
currentTimeMillis(), DateUtils.FORMAT_SHOW_TIME | DateUtils.FORMAT_SHOW_
DATE | DateUtils.FORMAT_ABBREV_ALL);
    efreshView.getLoadingLayoutProxy().setLastUpdatedLabel(label);
    // 获得加载布局，并更新
                    new GetDownTask(mPtflv).execute();
        }
        // 上拉加载启动异步
        @Override
        public void onPullUpToRefresh(PullToRefreshBase<ListView> refresh-
View) {
                    new GetUpTask(mPtflv).execute();
    // 新闻加载判断是否还有未显示数据
            } }
```

第十步：创建 GetDownTask 类并继承 AsyncTask<String, Void, Integer> 实现刷新方法。具体如代码 CORE0715 所示。

```java
代码 CORE0715 编写刷新方法
/* 此处添加实现刷新方法的代码 */
class GetDownTask extends AsyncTask<String, Void, Integer> {
            private PullToRefreshListView mPtrlv;
            public GetDownTask() {
            }
        public GetDownTask(PullToRefreshListView ptrlv) {
    // 重写 GetDownTask 方法，并传参
                    this.mPtrlv = ptrlv;
            }
    // 第三步执行 doInBackground
            @Override
            protected Integer doInBackground(String... params) {
                    return null;
            }
    / 第二步执行 onPostExecute
            @Override
```

```
protected void onPostExecute(Integer result) {
        super.onPostExecute(result);
        page = 1;
        arrayList.clear();
NetWork();                                // 刷新时访问网络
        mAdapter.notifyDataSetChanged();
        mPtrlv.onRefreshComplete();

} }
```

第十一步：创建 GetUpTask 类并继承 AsyncTask<String, Void, Integer> 实现加载方法。具体如代码 CORE0716 所示。

代码 CORE0716 编写加载方法

```
/* 此处添加实现加载方法的代码 */
class GetUpTask extends AsyncTask<String, Void, Integer> {
        private PullToRefreshListView mPtrlv;
        public String content = null;
        public GetUpTask() {
        }
        // 重写 GetUpTask 方法，并传参
        public GetUpTask(PullToRefreshListView ptrlv) {
                this.mPtrlv = ptrlv;
        }
// 第三步执行 doInBackground
        @Override
        protected Integer doInBackground(String... params) {
                return null;
        }
// 第二步执行 onPostExecute
        @Override
        protected void onPostExecute(Integer result) {
                super.onPostExecute(result);
                page = page + 1;                      // 加载操作时页码 +1
                if (NetUtils.isConnected(activity) == true) {
                        getSimulationNews();
                } else {
                        getLocalNews();                // 获取本地缓存
                }
```

项目七 无忧险 201

```
                    mAdapter.notifyDataSetChanged();
                    mPtrlv.onRefreshComplete();
        } }
```

第十二步：在无网络时执行的操作，通过 URL 路径访问本地缓存信息。具体如代码
CORE0717 所示。

代码 CORE0717 编写 URL 的使用方法

```
/* 此处添加通过 URL 路径访问信息的代码 */
public void getLocalNews() {
        String    url  =  URLCollect.BASE_URL  +  URLCollect.INSUR_URL  +
"page="+page+"&pagesize=10";                    // 声明 url 地址
        ArrayList<Insur> ret = new ArrayList<Insur>(); // 创建 ArrayList 对象
        String data = null;
        try {
// 将 URL 编程 String 类型的 key
String key = hashKeyForDisk(url);
// 获取本地信息
DiskLruCache.Snapshot snapShot = mDiskLruCache.get(key);
                if (snapShot != null) {
                        data = snapShot.getString(0);
                } else {
                        data= "[]";
                } } catch (IOException e) {
                e.printStackTrace();
        }
Gson gson = new Gson();                          // 创建 Gson 对象解析
        ArrayList<Insur> mList = gson.fromJson(data, new
TypeToken<List<Insur>>() {
        }.getType());
        for (Insur news : mList) {
                ret.add(news);
        }
        mAdapter.addNews(ret);                   // 添加到适配器
        mAdapter.notifyDataSetChanged();
    }
```

第十三步：给 Item 设置监听器，根据列表索引加载不同的信息。具体如代码 CORE0718
所示。

Android 模块化项目实战

代码 CORE0718 编写 Item 监听器

```
/* 此处添加实现 Item 点击的方法 */
public void onResume() {
        super.onResume();
        // 设定点击事件
        // 给列表的 Item 设置监听器
    mListView.setOnItemClickListener(new
AdapterView.OnItemClickListener() {
            @Override
    public void onItemClick(AdapterView<?> parent, View view, int position, long id) {
        int size = arrayList.size();    // 获得列表总数
        if (size != 0) {        // 如果列表信息不为空进行传参
// 跳转到 InsurItemActivity 界面
            Intent intent = new Intent(activity, InsurItemActivity.class);
            intent.putExtra("id", arrayList.get(position - 1).getInsur_id());
            intent.putExtra("title", arrayList.get(position - 1).getInsur_name());
            intent.putExtra("desc", arrayList.get(position - 1).getInsur_desc());
            intent.putExtra("content", arrayList.get(position - 1).getInsur_content());
            intent.putExtra("tel", arrayList.get(position - 1).getInsur_tel());
            startActivity(intent);
                } } }); }
```

第十四步：通过原生动作 ACTION_CALL 跳转到 InsurItemActivity 界面后实现打电话功能。具体如代码 CORE0719 所示。

代码 CORE0719 编写打电话功能

```
/* 此处添加实现打电话功能的方法 */
@Override
    public void onClick(View v) {
        switch (v.getId()) {            // 判断 id
        case R.id.iv_back:            // 关闭当前页面
            InsurItemActivity.this.finish();
            break;
    default:
        case R.id.iv_call:            // 进行打电话
            Intent intent = new Intent();
intent.setAction(Intent.ACTION_CALL);
// 系统默认的 action,用来打开默认的电话界面
```

项目七 无忧险 203

```
intent.setData(Uri.parse("tel:"+tel));        // 需要拨打的号码
InsurItemActivity.this.startActivity(intent); // 跳转至相应界面
break;
        } }
```

第十五步：为了保证文件的正确性和防止被盗用，最后要对这些文本进行加密处理。具体如代码 CORE0720 所示。

代码 CORE0720 编写加密方法

```
/* 此处添加加密文件的代码 */
public String hashKeyForDisk(String key) {
        String cacheKey;
        try {
        final MessageDigest mDigest = MessageDigest.getInstance("MD5");
                mDigest.update(key.getBytes());
                cacheKey = bytesToHexString(mDigest.digest());
        } catch (NoSuchAlgorithmException e) {
                cacheKey = String.valueOf(key.hashCode());
        } return cacheKey;
    }
    private String bytesToHexString(byte[] bytes) {
        StringBuilder sb = new StringBuilder();
for (int i = 0; i < bytes.length; i++) {        // 使用循环加密
                String hex = Integer.toHexString(0xFF & bytes[i]);
                if (hex.length() == 1) {
                        sb.append('0');
                } sb.append(hex);
        } return sb.toString();
    }
```

第十六步：运行项目，实现如图 7.2 所示效果。

本项目介绍了 U 酒保项目无忧险模块的实现。通过本项目的学习可以了解不同 UI 控件之间的区别，掌握 UI 更新方法，掌握上拉加载下拉刷新的使用方法。实现界面上拉加载，下拉刷新实时更新 UI 的功能。

pull	拉	refresh	刷新
datePicker	日期选择器	handler	处理者
looper	消息泵	thread	线程
user	用户	interface	界面
listener	收听者	updated	更新

一、选择题

1. 下列不属于 Android PullToRefresh 支持的控件是（　　　）。

A.ListView　　　　　　　　　　　B.GridView

C.SeekBar　　　　　　　　　　　D.WebView

2. 关于 PullToRefresh 叙述正确的选项是（　　　）。

A.Android PullToRefresh 是一个强大的拉动加载开源项目,支持各种控件刷新

B.Android PullToRefresh 是一个强大的拉动加载开源项目,支持少部分控件刷新

C.Android PullToRefresh 是一个强大的拉动刷新开源项目,支持少部分控件刷新

D.Android PullToRefresh 是一个强大的拉动刷新开源项目,支持各种控件刷新

3. 关于下拉刷新和上拉加载说法正确的是（　　　）。

A. 上拉加载的方法和下拉刷新的方法完全不同

B. 上拉加载的方法和下拉刷新的方法完全相同

C. 上拉加载的方法和下拉刷新的方法基本相同,不同点主要体现在设置刷新模式和绑定监听事件

D. 上拉加载的方法和下拉刷新的方法基本相同,不同点主要体现在设置刷新模式和绑定适配器

4. 下列关于 Handler 说法正确的是（　　　）。

A. 负责 Message 的发送及处理。使用 Handler 时,需要实现 handleMessage(Message msg) 方法来对特定的 Message 进行处理,例如更新 UI 等

B. 不断地从 MessageQueue 中抽取 Message 执行。因此,一个 MessageQueue 需要一个 Looper

C. 负责调度整个消息循环,即消息循环的执行场所

D. 其中包含了消息 ID,消息处理对象以及处理的数据等,由 MessageQueue 统一列队

5. 下列哪一个控件不属于 FrameLayout 体系（　　　）。

A.DatePicker　　　　　　　　　　B.RadioGroup

C.TimePicker　　　　　　　　　　D.CalendarView

二、填空题

1. 实现下拉刷新方法：____、____、____、____、____。
2. 上拉加载的方法和下拉刷新的方法不同点主要体现在_____和_____。
3. UI 控件的三要素：_____、_____、_____。
4. UI 更新的三种方法：_____、_____、_____。
5. Android UI 控件：_____、_____、_____、_____、_____、时钟控件、_____等。

三、上机题

编写代码实现下拉刷新功能。

项目八　服务器部署与报错处理

通过该项目的学习，了解 U 酒保项目服务器如何部署和使用，掌握 U 酒保项目的开发流程，学习解决 Android Studio 常遇报错的具体方法，具有解决 Android Studio 常遇报错的能力。在任务实现过程中：

- 了解 U 酒保项目服务器如何部署和使用。
- 掌握 U 酒保项目的开发流程。
- 掌握 Android Studio 常遇报错的具体解决方法。

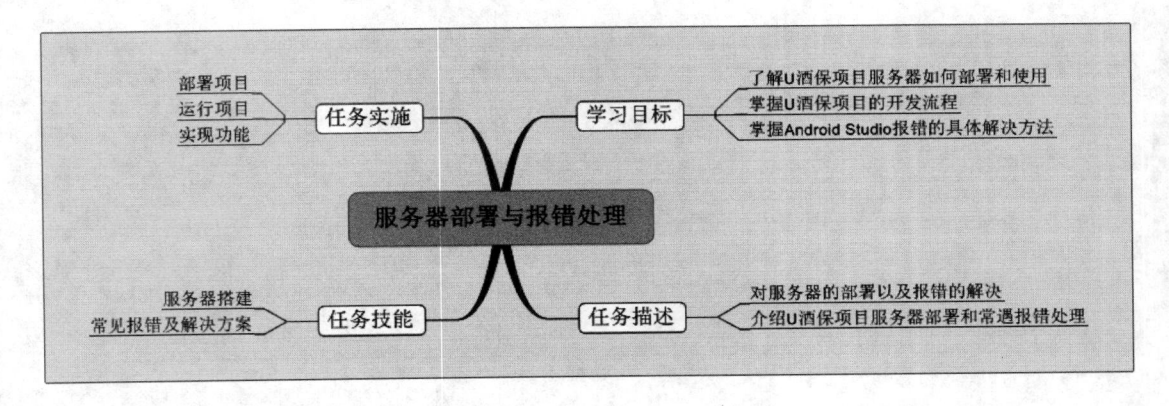

【情境导入】

U 酒保项目运行的最重要一环就是服务器的部署，服务器是网络中能对其他机器提供某些服务的计算机系统，U 酒保所需服务器的地方很多，例如：登录注册、健康助手界面、打车代驾界面、无忧险界面。在编写代码过程中可能会遇到很多种不同类型的错误问题，针对这些错误问题本项目将一一解决。

【功能描述】

本模块将介绍 U 酒保项目服务器端部署和常遇报错处理：

- 部署服务器。
- 解决服务器报错问题。
- 了解常遇报错。
- 常遇报错正确处理方式。

技能点 1　服务器搭建

1　服务器简介

服务器英文名称为"Server"，指在网络环境中为客户机（client）提供各种服务的特殊计算机。在网络中，服务器承担着数据的存储、转发、发布等关键任务，是各类基于客户机／服务器（C/S）模式网络中不可或缺的重要组成部分。

2　服务器特性

服务器特性是指在网络环境中，为客户提供各种服务，可以用服务器的四大特性衡量服务器的好坏，服务器的四大特性有可扩展性（Scalability）、易用性（Usability）、可用性（Availability）、可管理性（Manageability）。

（1）可扩展性

服务器必须具有一定的"可扩展性"，因为网络是多变的，特别是在当今信息时代。如果服务器没有一定的可扩展性，用户增多就不能胜任，一台造价高昂的服务器在短时间内就要遭到淘汰，这是任何企业都无法承受的。为了保持可扩展性，需要服务器具备一定的可扩展空间和冗余件。可扩展性体现在硬盘是否可扩充，CPU 是否可升级或扩展，系统是否支持 WindowsNT、Linux 或 UNIX 等多种可选主流操作系统等方面。

（2）易用性

服务器的功能相对于 PC 机复杂许多，不仅是硬件配置，更多的是软件系统配置。服务器要实现如此多的功能，没有全面的软件支持是无法实现的。因为软件系统增多，又可能造成服务器的使用性能下降，管理人员无法有效操纵。服务器的易使用性主要体现在服务器是否容易操作，用户导航系统是否完善，机箱设计是否人性化，有无关键恢复功能，是否有操作系统备份，以及有没有足够的培训支持等方面。

（3）可用性

服务器所面对的是整个网络的用户，而不是单个用户，要求服务器永不中断。在一些特殊应用领域，即使没有用户使用，服务器也得不间断地工作，因为它必须持续地为用户提供连接服务，不管是在上班，还是下班，也不管是工作日，还是节假日。这要求服务器必须具备极高的稳定性。一般服务器都要 7×24 小时不间断地工作，特别像一些大型的网络服务器。对于这些服务器来说，也许真正工作开机的次数只有一次，此后，它不间断地工作，一直到彻底报废。如果经常出问题，则网络不可能保持长久正常运作。为了确保服务器具有高的"可用性"，除了要求各配件质量过关外，还可采取必要的技术和配置措施，如硬件冗余、在线诊断等。

（4）易管理性

在服务器的主要特性中，还有一个重要特性，那就是服务器的"易管理性"。虽然服务器需要不间断地持续工作，但再好的产品都有可能出现故障。服务器虽然在稳定性方面有足够保障，但也应有必要的避免出错的措施，以及时发现问题，而且出了故障也能及时得到维护。这不仅可减少服务器出错的机会，同时还可提高服务器维护的效率。

3　服务器部署

在项目开发完成后，我们要部署项目，通常是将项目上传到云服务器中。具体分为以下三个步骤：

● 打包上传：将项目打包成 war 文件，然后将其传到远程服务器（在 MyEclipse 中直接将项目导出为 .war 文件）。

● 将 war 文件移动到 Tomcat 目录下的 webapps 下。

● 重启 Tomcat，访问我们的项目。

在这个过程中，我们需要注意。因为一般情况作为一个程序项目，肯定是有数据库的使用的。那么数据库部分怎么办呢？其实，只需要将我们已有的数据库转储为 sql 文件，然后将 sql 文件上传到云服务器上执行即可。以 mysql 为例，具体操作如下所示：

（1）转储为 sql 脚本，并上传。先在本地将我们项目使用的数据库转为 sql 文件，上传到云服务器上（可以利用 Navicat 将数据库转储为 .sql 文件）。

（2）执行 sql：然后进入 mysql 中执行该 sql 文件。（若服务器装有 Navicat，可直接用 Navicat 执行 .sql 文件，执行前需要选中存放表的数据库，应该与代码中数据库连接语句包含的数据库名保持一致）。

为了方便项目的测试，在编写的过程中我们所写的 U 酒保项目的服务器是通过本地的 Tomcat 实现的，把程序部署在 Tomcat 上，由 Tomcat 去运行，以达到所需的服务效果，在项目编写完成后，同样我们也要将其上传到云服务器。本地服务器的部署要使用到 MyEclipse，在成功配置相应环境变量之后，将服务项目"SpiritHelper"导入到 MyEclipse 当中。导入成功之后右键点击"SpiritHelper"，点击"Run As"，然后点击"MyEclipse server Application"，如图 8.1 所示。

项目八 服务器部署与报错处理 209

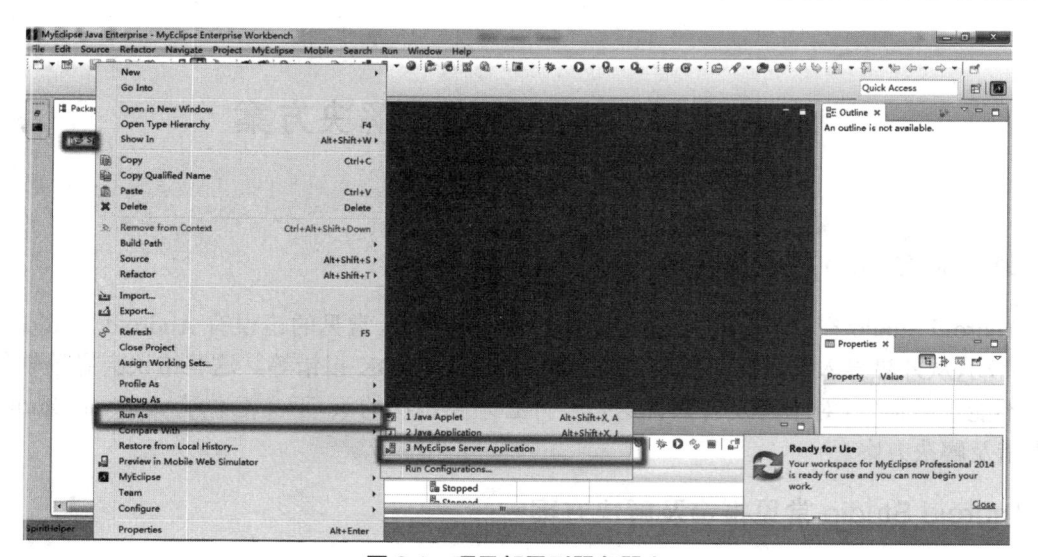

图 8.1 项目部署到服务器上

选择所需 Tomcat（现在 Tomcat 已经有多个版本，常用的是 Tomcat7.0 及以上版本，这里只列出了两种版本），如图 8.2 所示。

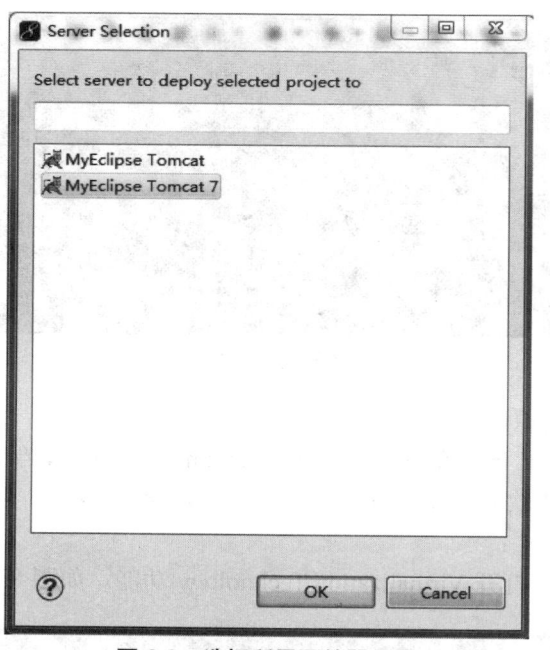

图 8.2 选择所需要的服务器

待服务项目成功在 Tomcat 上启动之后，U 酒保的服务器就部署完成了。用户可打开浏览器，在网址栏中输入"http://localhost:8080/SpiritHelper/jsp/login.jsp"进行测试。

技能点 2　常见报错及解决方案

1　Android Studio 报错简介

　　Android Studio 在开发过程中会出现各式各样的错误,常见的错误有 Android Studio 不能正常启动、从外界导入的程序不兼容、资源文件的丢失、SDK 报错等。这些错误会导致软件和程序不能正常运行,对于开发者来说这是一件必须要解决的事,下文汇总了 Android Studio 常见报错及解决方案。

2　Android Studio 常见报错及解决方案

　　(1)x86 emulation currently requires hardware acceleration
　　启动 Android 虚拟机时,出现如图 8.3 所示的错误对话框。

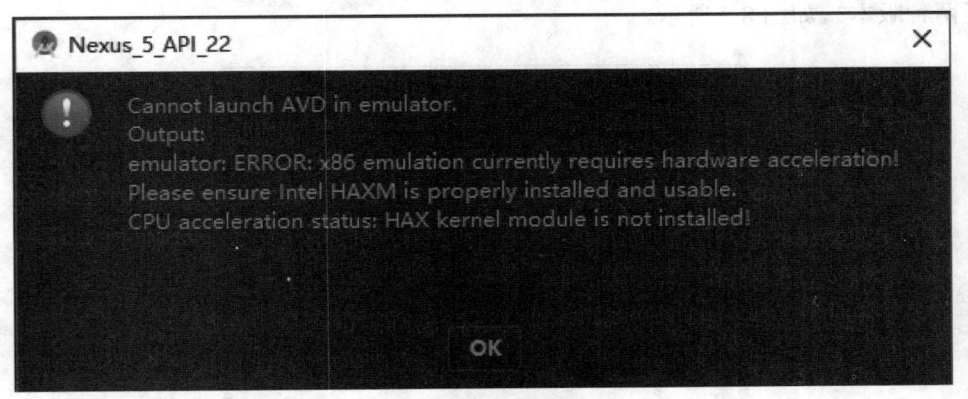

图 8.3　错误对话框

　　● 出现这种情况的原因:
　　x86 虚拟机是依赖于 Intel 的 Virtualization Technology 功能,当 Virtualization Technology 功能关闭或 Intel HAXM 软件未安装时,会导致模拟器启动失败。
　　● 解决该错误的方法:
　　在计算机的 BIOS 中打开 Virtualization Technology 功能。如图 8.4 所示。

图 8.4　BIOS 界面

- 安装 Intel HAXM。
- 成功启动 AVD 虚拟机。

（2）java.io.IOException: error=2

> Error: Cannot run program "/opt/android-sdk/build-tools/19.0.1/aapt": java.io.IOException: error=2, No such file or directory
> :Client:mergeDebugResources FAILED

- 出现这种情的况原因：

Android SDK 中的 adb、aapt 等程序是 32 位，当使用 Ubuntu64 位的系统时，Android Studio 就会报出该错误。

- 解决该错误的方法：

在 Ubuntu 64 位系统上安装 32 位兼容库。

（3）App 机器人位置（select run/debug Configuration）出现红叉

- 出现这种情况的原因：

因为文件换包导致了 Android 配置文件（AndroidManifest.xml）出现错乱。

- 解决该错误的方法：

先 clean 与 rebulde，如果不能解决，则进行下一步。

在 AndroidManifest.xml 文件中查看注册的 Activity 有没有报错。一般是清单文件错误问题，检查清单文件中 应用程序包名和 Activity 的名字。

（4）R 文件出现红色

出现如图 8.5 所示情况。

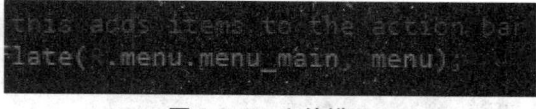

图 8.5　R 文件错误

- 出现这种情况的原因：

资源文件没有自动生成，缺少资源文件。

- 解决该错误的方法：

查看布局文件错误，直接进行编译，在 Message 面板根据给出的提示找出对应行的错误。

（5）v4 包的版本不一致

> java.lang.NoClassDefFoundError: com.jcodecraeer.devandroid.MainActivity

- 出现这种情况的原因：

不同的 module 组件中使用的 v4 包版本不一致。

- 解决该错误的方法：

在 build 中修改成一样的 v4 版本的包。

（6）出现如下所示错误

> Error:(26, 9) Attribute application@icon value=(@drawable/logo) from AndroidManifest.xml:26:9
>
> Error:(28, 9) Attribute application@theme value=(@style/ThemeActionBar) from AndroidManifest.xml:28:9
>
> is also present at XXXX-trunk:XXXXLib:unspecified:15:9 value=(@style/AppTheme)
> Suggestion: add 'tools:replace="android:theme"' to <application> element at AndroidManifest.xml:24:5 to override
>
> Error:Execution failed for task ':XXXX:processDebugManifest'.
> > Manifest merger failed with multiple errors, see logs

● 出现这种情况的原因：

Android Studio 的 Gradle 插件默认会启用 Manifest Merger Tool，若 Library 项目中也定义了与主项目相同的属性（例如默认生成的 android:icon 和 android:theme），则此时会合并失败，并报上面的错误。

● 解决该错误的方法：

在 Manifest.xml 的 application 标签下添加 tools:replace="android:icon, android:theme"（多个属性用"，"隔开，并且在 manifest 根标签上加入 xmlns:tools="http://schemas.android.com/tools"）。

或在 build.gradle 根标签上加上 useOldManifestMerger true。

（7）AndroidStudio SDK directory does not exists

出现如图 8.6 所示情况。

图 8.6 SDK 报错对话框

● 出现这种情况原因是：

因为网络上 Download 是一个开源的项目，如果把它导入 Android Studio 中，结果会使本来项目的 SDK 目录和下载项目的 SDK 目录不同，所以会出现这样的问题。

● 解决该错误的方法：

打开下载的项目根目录，找到 local.properties 文件，并打开，修改"sdk.dir"条目，改为系统下的 SDK 目录，如图 8.7 所示。

图 8.7 local.properties 文件

项目八 服务器部署与报错处理

（8）在编译的时候出现：Failure [INSTALL_FAILED_OLDER_SDK]

● 出现这种情况的原因：

Android Studio 自动设置了 compileSdkVersion。

● 解决该错误的方法：

方法一

修改 build.gradle 下的 compileSdkVersion "android-L" 为本机已经有的 SDK 版本，例如：compileSdkVersion 19。

方法二

打开 Project Structure，如图 8.8 所示。

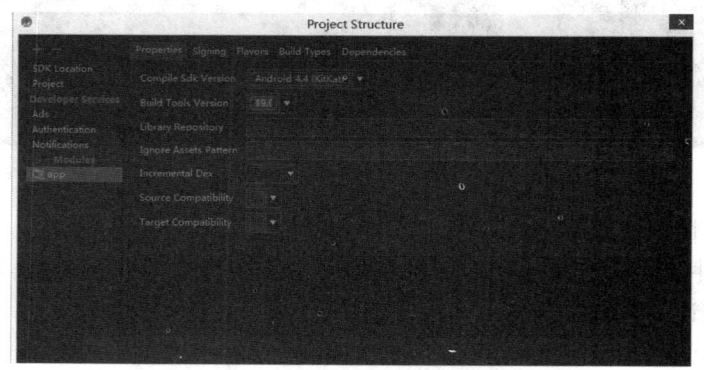

图 8.8　Project Structure 设置界面

把 Compile SDK Version 改为符合的版本，在 build.gradle 中 compileSdkVersion 会对应得到改变，如图 8.9 所示。

图 8.9　片段代码

修改 build.gradle 中 targetSdkVersion "L" 为上面设置的版本，如图 8.10 所示。

图 8.10　片段代码

（9）运行时出现 NullpointExcepition 异常

如图 8.11 所示。

图 8.11　空指针异常

● 出现这种情况的原因：

未定义的控件、返回空值、字符串变量未初始化、接口类型的对象没有用具体的类初始化等。

● 解决该错误的方法：

在 log 文件中找到相对应的行，有针对性的检查代码。

（10）数组越界异常

具体报错如下所示。

```
FATAL EXCEPTION: main
Process: package.xxx.xxx.xxx, PID: 10477
Java.lang.ArrayIndexOutOfBoundsException: length=3; index=3
    at Android.widget.AbsListView$RecycleBin.scrapActiveViews(AbsListView.java:6744)
    at android.widget.ListView.layoutChildren(ListView.java:1698)
    at android.widget.AbsListView.onLayout(AbsListView.java:2169)
    at android.view.View.layout(View.java:15794)
    at android.view.ViewGroup.layout(ViewGroup.java:5059)
    at android.widget.FrameLayout.layoutChildren(FrameLayout.java:579)
    at android.widget.FrameLayout.onLayout(FrameLayout.java:514)
    at android.view.View.layout(View.java:15794)
    at android.view.ViewGroup.layout(ViewGroup.java:5059)
    at android.widget.LinearLayout.setChildFrame(LinearLayout.java:1734)
    at android.widget.LinearLayout.layoutVertical(LinearLayout.java:1588)
    at android.widget.LinearLayout.onLayout(LinearLayout.java:1497)
    at android.view.View.layout(View.java:15794)
```

● 出现这种情况的原因：

例如，当使用 Listview 的多布局时，getItemViewType 需要从 0 开始计数，并且 getViewTypeCount 要大于 getItemViewType 中的数。

● 解决该错误的方法：

找出对应行的错误代码，修改至符合规则。

（11）添加第三方库出现的问题

```
Error:Execution failed for task ':app:processDebugManifest'.
Manifest merger failed :
 uses-sdk:minSdkVersion 14 cannot be smaller than version 19 declared in library [com.
github.meiko
z:basic:2.0.3]
/AndroidStudioCode/EnjoyLife/app/build/intermediates/exploded-aar/
com.github.meikoz/basic/2.0.3/AndroidManifest.xml
Suggestion: use tools:overrideLibrary="com.android.core" to force usage
```

● 出现这种情况的原因：

引入的第三方库最低支持版本高于项目的最低支持版本，异常中的信息显示：项目的最低支持版本为 14，而第三方库的最低支持版本为 19，所以抛出了这个异常。

● 解决该错误的方法：

在 AndroidManifest.xml 文件中标签中添加：<uses-sdk tools:overrideLibrary="xxx.xxx.xxx"/>，其中的 xxx.xxx.xxx 为第三方库包名，如果存在多个库有此异常，则用逗号分割它们，例如：<uses-sdk tools:overrideLibrary="xxx.xxx.aaa, xxx.xxx.bbb"/>。

（12）Need XXX permission 报错

报错信息如图 8.12 所示。

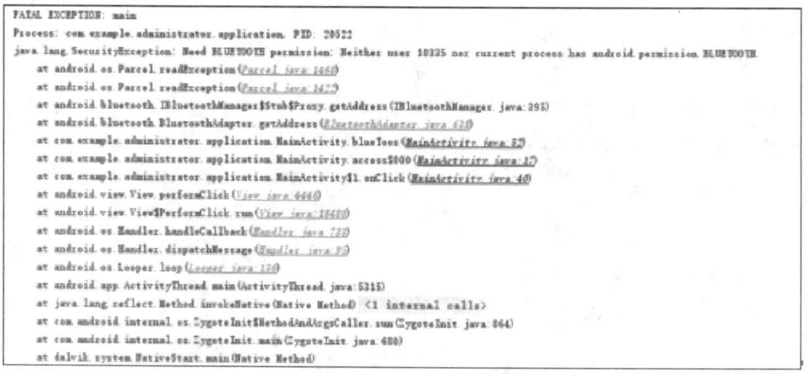

```
FATAL EXCEPTION: main
Process: com.example.administrator.application, PID: 20522
java.lang.SecurityException: Need BLUETOOTH permission: Neither user 10335 nor current process has android.permission.BLUETOOTH.
    at android.os.Parcel.readException(Parcel.java:1468)
    at android.os.Parcel.readException(Parcel.java:1422)
    at android.bluetooth.IBluetoothManager$Stub$Proxy.getAddress(IBluetoothManager.java:293)
    at android.bluetooth.BluetoothAdapter.getAddress(BluetoothAdapter.java:638)
    at com.example.administrator.application.MainActivity.blueTees(MainActivity.java:52)
    at com.example.administrator.application.MainActivity.access$000(MainActivity.java:17)
    at com.example.administrator.application.MainActivity$1.onClick(MainActivity.java:40)
    at android.view.View.performClick(View.java:4446)
    at android.view.View$PerformClick.run(View.java:18420)
    at android.os.Handler.handleCallback(Handler.java:733)
    at android.os.Handler.dispatchMessage(Handler.java:95)
    at android.os.Looper.loop(Looper.java:136)
    at android.app.ActivityThread.main(ActivityThread.java:5315)
    at java.lang.reflect.Method.invokeNative(Native Method) <1 internal calls>
    at com.android.internal.os.ZygoteInit$MethodAndArgsCaller.run(ZygoteInit.java:864)
    at com.android.internal.os.ZygoteInit.main(ZygoteInit.java:680)
    at dalvik.system.NativeStart.main(Native Method)
```

图 8.12 缺少蓝牙权限

● 出现这种情况的原因：

缺少蓝牙权限。

● 解决该错误的方法：

添加蓝牙权限，如图 8.13 所示。

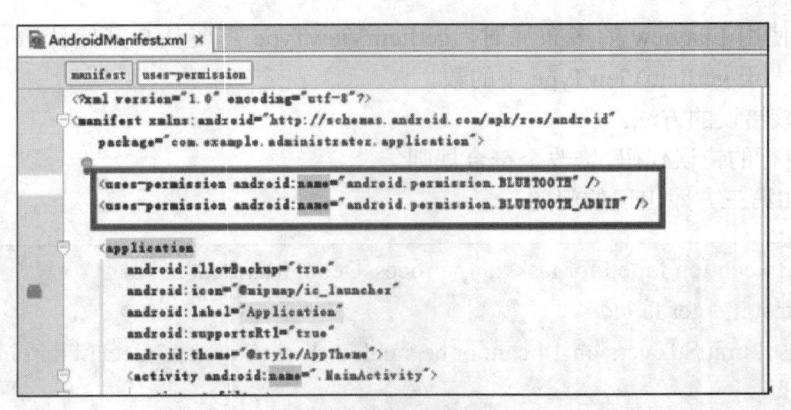

图 8.13　添加蓝牙权限

（13）JSON 数据解析问题

异常信息如下所示：

> org.json.JSONException: Value of type java.lang.String cannot be converted to JSON-Object

- 出现这种情况的原因：

JSON 串头部出现字符："\ufeff"。

- 解决该错误的方法：

```
/* 异常信息: org.json.JSONException: Value  of type java.lang.String cannot be con-
verted to JSONObject
*JSON 串头部出现字符:"\ufeff" 解决方法
* @param data
* @return
*/
public static final String removeBOM(String data) {
if (TextUtils.isEmpty(data)) {
    return data;
}
if (data.startsWith("\ufeff")) {
    return data.substring(1);
    }
else {
return data;
    } }
```

（14）not found ndk()

异常信息如下所示：

项目八 服务器部署与报错处理 217

Error:(15, 0) Gradle DSL method not found: 'ndk()' method-not-found-ndk

● 出现这种情况的原因：
由于 ndk() 配置在 build.gradle 配置文件中位置出错导致的。
● 解决该错误的方法：

```
apply plugin: 'com.android.application'
android {
    compileSdkVersion 23
    buildToolsVersion "23.0.2"
    defaultConfig {
        applicationId "com.guitarv.www.ndktest"
        minSdkVersion 17
        targetSdkVersion 23
        versionCode 1
        versionName "1.0"
// 修改 build.gradle 中 ndk 的配置位置
        ndk {
            moduleName = "HelloJNI"
        } sourceSets.main {
            jni.srcDirs = []
            jniLibs.srcDir "src/main/libs"
        } }
    buildTypes {
        release {
            minifyEnabled false
            proguardFiles getDefaultProguardFile('proguard-android.txt'), 'proguard-rules.
pro'
        } } }
```

拓展 怎么样？上边列出的在 Android Studio 开发过程遇到的错误够不
够详细？解决方案够不够全面？什么？还是不够？别着急，就知道你们是爱
学习、爱专研的好学生。我们的小码已经迫不及待的等大家去扫了。还不快
快行动？

通过以下步骤，实现 U 酒保项目的部署与实现。

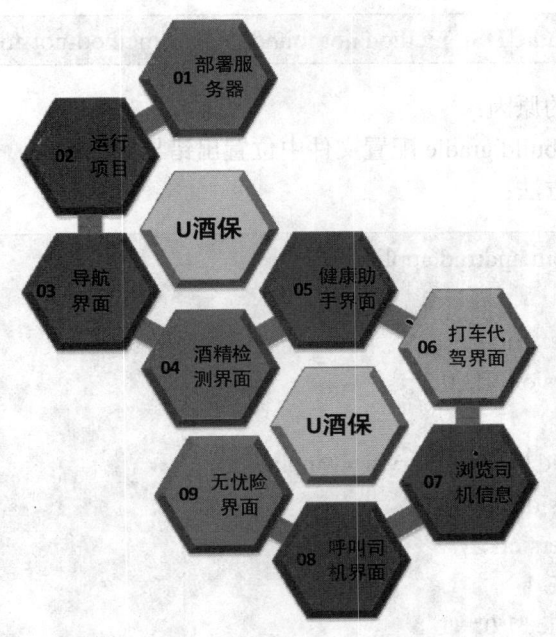

具体步骤如下所示。

第一步:服务器部署完成,运行完整项目,出现如图8.14所示效果。

图8.14 闪屏界面

第二步:导航界面引导功能,在用户首次下载并安装该软件时,会进入导航界面提示用户当前版本是否需要更新,如果无需更新则进入欢迎界面。效果如图8.15所示。

第三步:安全自测中酒精检测模块,在酒精检测模块已连接的情况下,通过单击"立即检测"按钮进行检测,得到相应酒精浓度,如果检测模块未连接或连接不成功,将提示用户重新连接。效果如图8.16所示。

第四步:健康助手模块,将用户一段时间内检测的酒精浓度进行记录并通过折线图显示。效果如图8.17所示。

项目八 服务器部署与报错处理

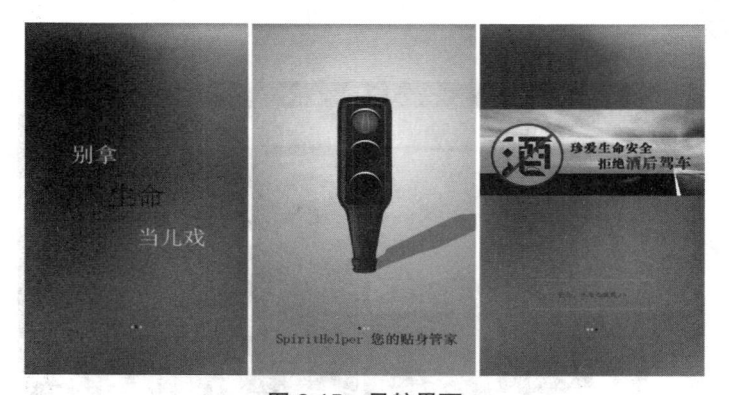

图 8.15 导航界面

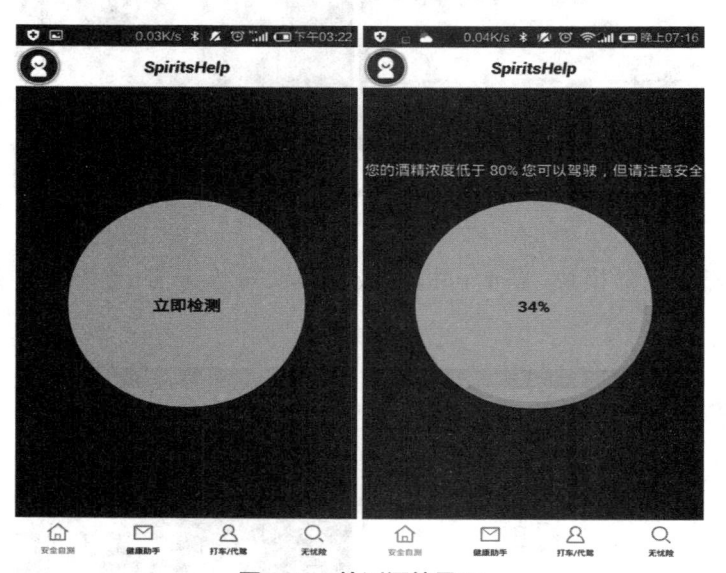

图 8.16 检测酒精界面

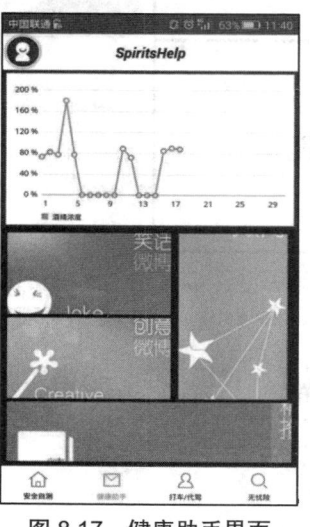

图 8.17 健康助手界面

第五步：健康助手模块科普知识功能，分为养生知识、驾车技巧、酒驾危害、自救常识四类。效果如图 8.18 所示。

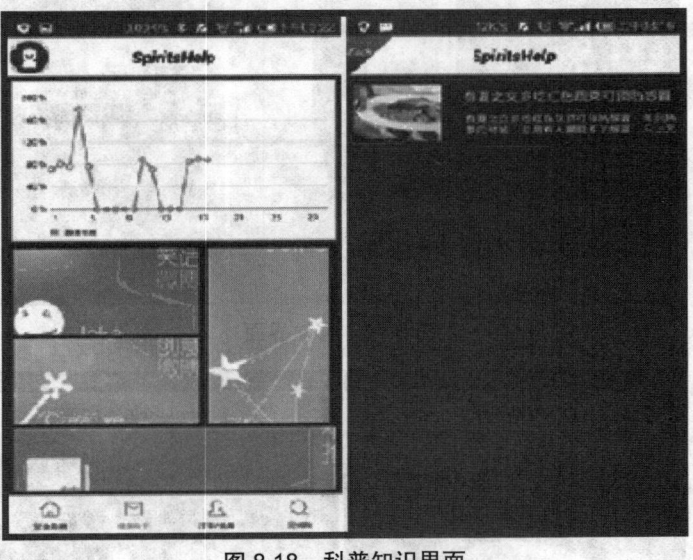

图 8.18　科普知识界面

第六步：打车/代驾模块，显示司机等级、驾龄、司机编号、姓名等详细信息。效果如图 8.19 所示。

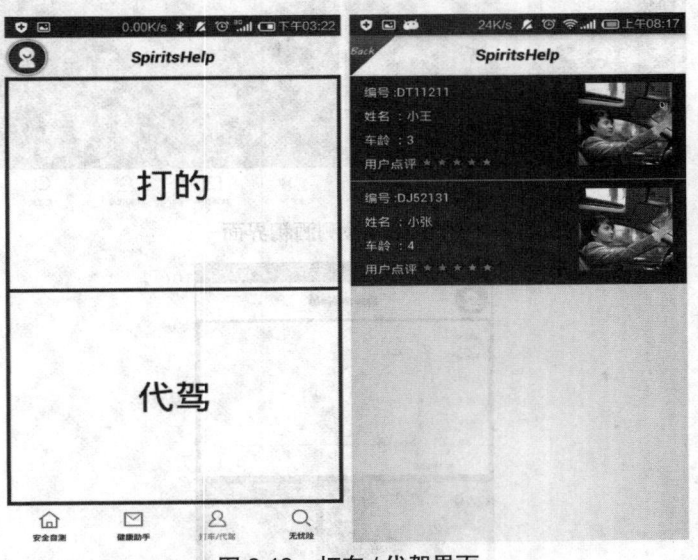

图 8.19　打车/代驾界面

第七步：浏览司机信息并实现呼叫司机功能。效果如图 8.20 所示。

项目八　服务器部署与报错处理　　221

图 8.20　呼叫服务界面

　　第八步：无忧险模块，保险推广功能可查看保险信息并调用系统电话服务进行咨询。效果如图 8.21 所示。

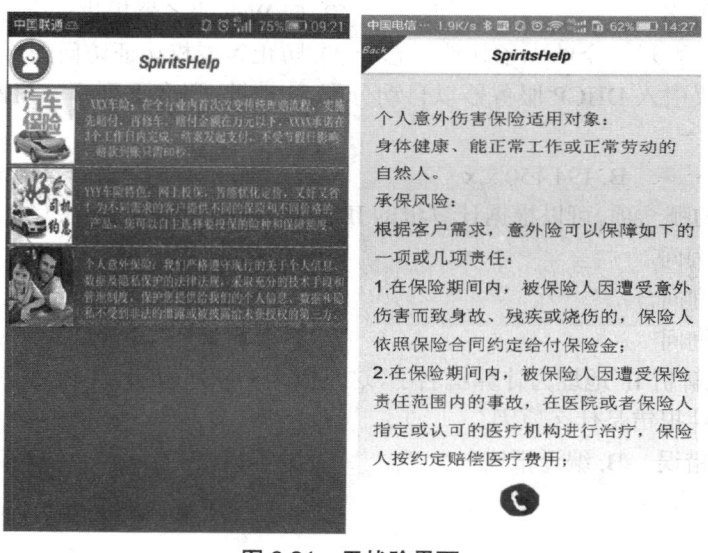

图 8.21　无忧险界面

　　本项目介绍了 U 酒保服务器的部署与运用及 Android Studio 常遇错误处理，通过对本项目的学习可以了解服务器的部署，对 Android Studio 常遇错误正确处理，重点掌握 Android Studio 错误解决方式。达到遇到程序错误不慌张，能够自己单独解决程序错误问题的程度。

reliability　　　　可靠性　　　　availability　　　　可扩展性

scalability　　　　方式　　　　　usability　　　　易用性

manageability　　 易管理　　　　properties　　　　性能

structure　　　　 结构　　　　　compile　　　　　编译

technology　　　　技术　　　　　module　　　　　模块,组件

一、选择题

1. 下面不是服务器特征的是（　　　）。

A. 可靠性　　　　　B. 易用性　　　　　C. 多态性　　　　　D. 可拓展性

2. HTTPS 可以允许用户实现接下来的哪一个操作（　　　）。

A. 建立数字签名　　　　　　　　　B. 向 Web 服务器提供安全连接

C. 通过局域网发送安全的电子邮件信息　　D. 防止为授权认证访问局域网

3. 如果您提议引入 DHCP 服务器以自动分配 IP 地址,那么下列哪一组网络 ID 将是您最好的选择（　　　）。

A. 24.x.x.x　　　　B. 194.150.x.x　　　　C. 172.16.x.x　　　　D. 206.100.x.x

4. 使用下列的哪一项,可以根据计算机的 IP 地址获得计算机的名字（　　　）。

A. 正向查找询问

B. 反向查找询问

C. 向后查找询问

D. DNS 不能解析 IP 地址为计算机名字,这是因为 DNS 使用计算机名字作为索引

5. 出现 R 文件报错是什么原因（　　　）。

A. 布局文件错误　　B. 编译报错　　　　C. 版本错误　　　　D. 缺少资源文件

二、填空题

1. 服务器是计算机的一种,是网络中为客户端计算机提供各种服务的高性能的计算机,服务器英文名称为＿＿＿＿＿＿。

2. 电子邮件服务器使用＿＿＿＿＿＿协议向外发送电子邮件。

3. 路径错误将导致编写好项目报＿＿＿＿＿＿错。

三、简答题

1. 简述服务器特征。

2. 简述 R 文件报错解决方法。